Lime and Alternative Binders

in East Africa

Lime and Alternative Binders
in East Africa

Edited by

Elijah Agevi, Otto Ruskulis
and Theo Schilderman

Intermediate Technology Publications
in association with the
Overseas Development Administration
1995

Intermediate Technology Publications
103-105 Southampton Row
London WC1B 4HH, UK

© Intermediate Technology Publications 1995

A CIP record for this book is available from the British Library

ISBN 1 85339 330 4

Artwork by DesignWrite Productions
Printed in UK

Acknowledgements

This book is the published proceedings of a workshop on Lime and Alternative Cements in East Africa, which took place at Tororo in Uganda in December 1994.

The workshop and this publication would not have been possible without the financial support of the Overseas Development Administration (ODA) of the British Government, and this support is gratefully acknowledged. In particular, ODA funded a series of national surveys on the cements sector in East Africa, which have been revised for this publication, and which formed the basis of discussion at the workshop. This in turn led to the formulation of resolutions, recommendations and an action plan to take the development of the lime and alternative binders sub-sector forward in East Africa.

The views expressed in this publication and the recommendations proposed do not necessarily coincide with those of ODA.

Abbreviations and acronyms

Technical:

BS	British Standards
CIS	Corrugated iron sheet
LPC	Lime-pozzolana cement
LOI	Loss on ignition
MJ	Megajoule (a unit of energy)
MPa	Megapascal (a unit of stress, pressure or strength)
OPC	Ordinary Portland cement
PFA	Pulverized fuel ash
PPC	Portland-pozzolana cement
QAS	Quality assurance system
R&D	Research and development
RHA	Rice husk ash
RHAC	Rice husk ash cement
VSK	Vertical shaft kiln

Development-related:

CBO	Community-based organization
ESAP	Economic Structural Adjustment Programme
NGO	Non-government organization
PTA	Preferential trade area for eastern and southern Africa

Institutions:

BASIN	Building Advisory Service and Information Network
BRU	Building Research Unit (Tanzania)
CAS	Cements and Binders Advisory Service
GTZ	German Technical Co-operation
HABRI	Housing and Building Research Institute (Kenya)
IDRC	International Development Research Centre
ILO	International Labour Organisation
IT	Intermediate Technology
ITDG	Intermediate Technology Development Group
MLHUD	Ministry of Lands, Housing and Urban Development (Uganda)
NCC	National Construction Council (Tanzania)
NRM	National Resistance Movement
ODA	Overseas Development Administration (of the British Government)
SIDO	Small Industries Development Organization (Tanzania)
STCDA	Stone Town Conservation and Development Authority (Zanzibar)
UCI	Uganda Cement Industry
UNBS	Uganda National Bureau of Standards
UNCHS	United Nations Centre for Human Settlements
UNDP	United Nations Development Programme
UNECA	United Nations Economic Commission for Africa
UNICEF	United Nations International Children's Emergency Fund
UNIDO	United Nations Industrial Development Organization
UON	University of Nairobi (Kenya)
URC	Uganda Railways Corporation

Conversion rates used in this document

The following approximate currency conversion rates
to one US$, current at June 1994, were used:
500 Tanzanian Shillings (TSh)
420 Sudanese pounds (Ls)
1,000 Ugandan Shillings (USh)
50 Kenya Shillings (KSh)

CONTENTS

Acknowledgements		v
Abbreviations and acronyms		vi
INTRODUCTION		viii
SECTION 1	**Production and consumption of binders in East Africa**	1
	Tanzania	3
	Uganda	31
	Kenya	43
	Zimbabwe	60
SECTION 2	**Lime production: Case studies**	63
	Nyalakoti Farming and Lime Works, Tororo, Uganda	65
	Experimental lime kiln at Tororo, Uganda	68
	Kigezi Twimukye Lime, Kisoro, Uganda	70
	Equator Lime, Kasese, Uganda	72
	Hima Lime Works, Uganda	75
	Homa Lime Company, Kenya	79
	Kenya Calcium Products, Tiwi, Kenya	83
	Production of lime in Zanzibar	85
	Wachumico Co-operative Society, Tanzania	90
	Kigamboni Lime Works, Tanzania	93
	Amboni, Tanga District, Tanzania	97
	Village Lime Kilns: The SIDO experience in Tanzania	99
	Kassala area (Al Gira), Sudan	101
	The lime industry in Malawi	104
SECTION 3	**Lime-pozzolana**	108
	HABRI's experience with pozzolanas in Kenya	115
	Pozzolanic cements in Uganda	121
	Research of the Building Research Unit (BRU), Tanzania	124
SECTION 4	**The use of binders in low-income housing**	126
	Kenya	128
	Uganda	144
	Tanzania	152
	Zanzibar	161
SECTION 5	**Workshop resolutions and recommendations**	163

INTRODUCTION

The East African region is in the process of major change both socially and economically. Towns and cities are expanding very quickly as the population increases, and more and more people are moving from rural to urban areas. In towns and cities people have different expectations of housing from those in rural areas. The house has to be permanent, it should not need a lot of time spent maintaining it, it should be secure from intruders and not present a fire risk. As a result, materials such as Ordinary Portland Cement (OPC) have been in great demand causing shortages and, in places, high prices. Shanty towns have developed around the larger cities as poorer people, who are unable to afford the high price of OPC, build with whatever they can afford or salvage.

In Tanzania, Kenya and Uganda the demand for OPC is forecast to increase rapidly over the next decade while installed capacity for OPC production is forecast to increase only slowly, so existing shortages can only get worse. OPC is already being imported, especially to Uganda, and this necessitates spending scarce foreign exchange. Lime is produced in all three countries to some extent, but only in Tanzania can it be considered to be a significant economic activity. There is also considerable potential to produce lime-pozzolana cement using volcanic ash, rice husk ash or pulverized burnt clay as the active pozzolana. Pozzolanas are in fact already being added to OPC to extend the cement.

It is also recognised that many building applications do not require the high strength of OPC, and lime or lime-pozzolana cement could make a completely satisfactory substitute in such cases. Moreover these so-called alternative cements can be produced at a small scale, at a fraction of the cost of OPC and for a local market, so their production level can be much more responsive to local demands than that of OPC, which is produced in a few large centralized plants far from many of its markets.

It is therefore important to look at what alternative cements could offer in lessening current and forecast cement shortages in East Africa. It is also important to look at the constraints lime producers are currently facing when consideration is given to increased lime production. These constraints are many and diverse and it is evident that they greatly inhibit the development of the lime industry in East Africa.

These circumstances formed the background to a Regional Workshop on Lime and Alternative Binders organized by the Intermediate Technology Development Group (ITDG) in conjunction with the Government of Uganda. The workshop was held in Tororo, Uganda in December 1994. It was attended by about 50 participants from six African countries: Tanzania, Uganda, Kenya, Malawi, Sudan and Zimbabwe; from the UK; and from two international agencies: Unido and Habitat. The participants represented various interests in the binders sector and consisted of producers, users, researchers, academics, NGOs, donors and Governments. The workshop was officially opened by the Ugandan Government Minister of State for Lands, Housing and Urban Development while it was closed by the Permanent Secretary from the same ministry. The papers which were presented at the workshop consisted of a series of national surveys of production and use of binders, a series of case studies on individual lime production plants in the region and some papers on research on lime-pozzolana cement in the area, including details on pilot projects to introduce production of this type of cement.

The workshop was concerned about the prevalent lack of adequate and affordable shelter for the majority of the region's population. The high cost of building materials, particularly cement, was noted as a major contributing factor to this problem, whereas unnecessarily high and restrictive standards and regulations do play a role as well. Lime and other binders, which could replace cement, are produced in the region on a small scale; they need to be further popularized, and producers as well as users made aware of more appropriate technologies for their production and application. The workshop noted with particular concern that most producers of such materials do make excessive and inefficient use of fuelwood, which has a serious impact on the environment, and that urgent steps are needed to improve this situation.

The workshop listed a substantial number of binders that could be alternatives to Portland cement, including lime, natural and artificial pozzolanas (when activated by lime or cement), various types of soil, cow dung and ash, liquor residues and molasses, gypsum and animal glues. Amongst these, lime is currently the predominant alternative binder in the region, and a substantial part of the discussion focused on its production and application, as will appear from the report below. The other alternatives are often insufficiently known or explored and would, in the first place, require substantial additional research.

An important element of the workshop was discussion sessions on the current binders situation in East Africa, and the role alternative cements could play in the future. The discussions initially took place in three groups focused on producers, users and policy makers respectively. Each group identified the main problems within their fields of the binders sector and at the final plenary session these were condensed to a total of 13 main issue/problem areas, and these were deliberated upon to produce an overall set of resolutions and an action plan to resolve or minimize these problem areas.

SECTION 1

Production and consumption of binders in East Africa

Summary

Binders are essential construction materials. Their function of binding other materials together is fundamental to building and few, if any, building projects today can do without them. In the formal sector, the availability of binders like cement is an essential prerequisite for capital formation, and the per capita consumption of cement is often considered an important indicator of the economic progress of a country. In the informal sector, largely composed of low-income housing, commercial binders have been less used traditionally, but if the quality of this housing is to improve over the years, binders are again essential as a means to protect or stabilize the earth of which these houses are so often built, and for other building elements. The informal sector, however, may not necessarily require the same type and quality of binder as the formal sector.

The production of binders is also important in terms of national development: the choice of the most appropriate materials and technologies can have a major impact on employment and income generation, capital and foreign exchange requirements, the types and quantities of energy used, the environment and transport requirements.

This section looks at the production and consumption of binders in East Africa. The main binder in this context is Ordinary Portland Cement (OPC), but lime is another important one. Moreover, pozzolanas are slowly finding their way into binder production, in combination with cement and with lime. The section on Kenya in particular pays attention to the consumption of traditional binders, as they are used in low-income housing in the whole region.

The material presented in this chapter has been assembled from in-depth studies by researchers in Tanzania, Uganda and Kenya, who have compiled available documentation, visited various producers, looked at the trade and talked to users. The sections on these countries are only a summary of the available data. At the end of the chapter a brief section on Zimbabwe in Southern Africa provides an interesting comparison.

There are great differences between the three East African countries in terms of binder production and consumption, as follows.

Kenya has two large cement plants which operate very efficiently; current cement consumption stands at about 40 kg/capita/year, which is down from 56 kg in 1991. The lime sector is relatively underdeveloped, with two large companies producing at less than half their capacity and very few artisanal producers. The country is moving towards adding more pozzolanas to cement, effectively producing a Portland-Pozzolana Cement (PPC); a few blended cements are produced, and there are attempts to produce lime-pozzolanas. There is also an interest in the use of more traditional binders such as special earths or cow-dung and ash which is less pronounced, though present, in the other countries.

Tanzania has three large cement plants, with a total capacity very close to that of Kenya, but they have a history of much lower capacity utilization. One has now been overhauled and operates well, and a second is in the process; the same is needed quite badly in the third factory. Cement consumption has been static at about 22 kg/capita/year, or about half that of Kenya, for several years now. But lime production and consumption in Tanzania is highest in the region, around 2 kg/capita/year. The bulk of that lime is produced from heap burning; larger-scale production is often facing problems. An attempt to produce lime-pozzolana, around 15 years ago, never really reached the market, but the production of PPC is now under way in one cement factory.

Uganda has been through a difficult period in the 1970s and 80s. Its two cement factories are run down. One is barely producing, the other one gradually improving but nowhere near a reasonable output yet. As a result, most cement consumed in Uganda is imported, largely from Kenya and Tanzania; current consumption stands at 10.5 kg/capita/year, only half of that in Tanzania, but real demand is stated to be at least 15 kg. In terms of lime production and consumption, Uganda is in between Kenya and Tanzania, with a fair number of intermittent kilns. Lime-pozzolana production is increasing in a few locations.

Zimbabwe, in comparison, comes closest to Kenya; it has three large cement factories, with a total capacity close to those of Tanzania and Kenya, which hardly cover local demand; but cement consumption is a lot higher, at 115 kg/capita/year. This may be some indicator for a potential growth in demand in East Africa in years to come. There is also a fairly high demand for lime, particularly in industry, but there are relatively few producers and a lot of lime is imported. There is great scope for intermediate scale lime production in Zimbabwe.

Notwithstanding the differences in binder production and consumption between the East African countries, there are also a number of issues they have in common:

Energy is the most common problem: its cost is often nearly half the production cost of cement or lime. The cement industry and the large-scale lime plants mainly use fuel oil which has to be imported. Whereas fuel consumption in the cement industry is typically 4-5 MJ/kg cement, the range is a lot larger in lime production, from about 5 MJ/kg lime for oil-fired kilns, to 15-30 MJ/kg for intermittent kilns, and even 35 MJ/kg for heap burning. Fuelwood is the main source of energy for lime burning, and it is estimated that Tanzania loses about 100,000 trees per year for that purpose, with all the negative consequences for the environment. More fuel-efficient lime production is a high priority. The introduction of pozzolanas which generally do not need fuel for firing can help to reduce the fuel consumption of binders considerably, but that depends to some extent on the amount of milling they require.

Transport is another problem area; binders are bulky and heavy and therefore it makes sense to produce them as close to markets as possible. It is known that having to transport cement to the more outlying areas of Tanzania can more than triple its price to the consumer. Currently, binders are mostly transported by road, because the East African rail network simply does not have the required capacity. Unless that is resolved, it would make sense to produce binders in a much more decentralized way in the future; this could include mini-cement plants, as suggested in the Tanzanian case, as well as lime or lime-pozzolana production units.

Employment is an increasing issue in the region; the production of lime creates 5-10 times more jobs per tonne produced than cement production, and it does so at only a fraction of the investment cost per job. From this perspective, countries should positively stimulate the production of alternative binders such as lime.

The affordability of binders is another issue. A bag of cement costs 40% of a monthly minimum wage in some areas of the region. The cost of lime, as that of lime-pozzolana, is generally lower, sometimes not more than half the price of cement. This helps, but it remains relatively expensive for the lowest income groups. As suggested in the section on Kenya, traditional binders therefore deserve some more attention, and ways to upgrade them may have to be found.

Pozzolanas are being used increasingly in each country, although not on any great scale yet. There is some reluctance, in particular towards the use of lime-pozzolanas, which is probably more innovative than using PPC. But pozzolanas deserve to be used more because of their potential to save energy, to make use of various waste materials such as rice husks or bagasse, and to make limestone reserves last longer.

—Theo Schilderman

Tanzania

Boniface Muhegi and Theo Schilderman

Raw Materials

Requirements
The production of Portland cement requires calcium carbonate, usually in the form of limestone, other minerals containing silica, alumina and iron, commonly in the form of clays, silt or shale, and a small amount of gypsum. It also needs fuel, in the form of coal or oil, and a good supply of electricity.

Lime is produced from calcium carbonate, usually limestone but occasionally coral, chalk or seashells, using a fuel, which traditionally is firewood, but coal, charcoal and oil are other options; the larger lime kilns in Tanzania in fact do use heavy fuel oil, and most of the smaller ones wood, except for a few in Mbeya region, which use the local coal (Ikomba, 1994b).

Pozzolanas are not binders in their own right, but can be activated by lime or cement; the resulting binders are lime-pozzolanas or Portland Pozzolana Cements, or hybrid forms of those. Mbeya Cement Company has recently started utilizing the natural pozzolanas in Mbeya to produce PPC. Pozzolanas come in various types; there are natural pozzolanas, in the case of Tanzania particularly volcanic ash, and artificial ones, including fired clays, rice husk ash and bagasse ash, as well as fly ash from coal burning.

Limestone and coral
In Tanzania, there are over 250 known deposits of carbonatic rocks, occurring in all regions; most of those are given on Map 1. These include mostly dolomitic marbles and dolomites, too high in magnesia for most applications, Jurassic-Quaternary limestones of the Coastal area as well as redeposited carbonates (travertine, calcrete etc.), which are currently exploited, and limestones of the Karoo, about which little is known but which may have potential (Austroplan, 1990). There are major occurrences in Arusha, Coast, Dodoma, Iringa, Kilimanjaro, Lindi, Mara, Mbeya, Morogoro and Tanga regions (UNECA, 1983). According to Austroplan (1990), there are proven reserves of 25 million tonnes of Pleistocene coral limestone at Wazo Hill, 20 km north of Dar es Salaam, exploited by the Tanzania Portland Cement Company (TPCC), 47 million tonnes of Jurassic limestone at Tanga, used by the Tanga Cement Company (TCC), and 50 million tonnes of travertine at Songwe, SE of Mbeya, exploited by the Mbeya Cement Company (MCC). There are also numerous deposits in Zanzibar, but these have not been extensively surveyed, except for one at Dimani, 14-15 km south of Zanzibar town, where the reserves are estimated at 20 million tonnes of high purity limestone (Suleiman, 1994).

Zanzibar Island and Pemba Island are largely made up of coral, and the same material also occurs on the coastal belt of Tanzania; in all those areas it is extensively exploited as a source of calcium carbonate. Along the coast, coral is regularly broken from the reefs which in the long term could lead to coastal erosion (Ministry of Tourism, Natural Resources and Environment, 1991). The large reserves of limestone in the hands of the cement industry are generally of good quality. Samples of the Tanga limestone are found to have a CaO content of between 52 and 55% and very few impurities; whereas the same applies to the Songwe travertines; the quality of limestone at Wazo Hill is somewhat more varied, with CaO contents ranging between 45 and 54% and occasionally more contamination by clay minerals (ranging from 3.3 to 16.5% overall). The quality of the smaller reserves can vary, but many of the deposits are suitable for lime production. The deposits used by Tanga Lime are as good as those used by the cement factory, and a sample of limestone from the Mji Mwema plant just South of Dar es Salaam also had a high CaO content; but samples from the Mandawa basin between Dar es Salaam and Lindi were more variable, with CaO contents ranging between 44 and 54%, and up to 21% of clay minerals. (Austroplan, 1990). In fact, dozens of those deposits are currently being exploited by small- to medium-scale lime producers, mainly in the coastal zone, on the islands, and along the Iringa-Dodoma-Arusha-Kilimanjaro axis.

The deposits claimed by the three cement plants have sufficient reserves of limestone for decades of future production. As to the smaller deposits, there is less information about their reserves but, given the current and foreseeable levels of lime production, this raw material is unlikely to become scarce for the foreseeable future.

Production plants of cementitious binders are best located near to sources of limestone, since that is the principal component of the binder, and it loses weight during the production process. The accessibility of limestone sites is therefore an important consideration in deciding on the best locations of production units; sites need to be easily accessible for the other raw materials, and the final product needs transport to markets. Tanzania is a fairly large country, with a relatively low population density, and a low level of infrastructure, which also lacks maintenance. In the past, its industry had been plagued by supply problems of various types, which for instance have contributed to the low capacity utilization of cement plants mentioned above, as well as by a lack of transport capacity for its products; as a result, many outlying areas have experienced scarcities of binders. It may be expected that the current liberalization of the economy could help, at least partly, to overcome these problems, but at a price. It therefore would make sense, in the national interest, to consider a more decentralized net of binder producers, by making better use of the large numbers of scattered limestone deposits in the country.

Entrepreneurs have to obtain a mining licence for the quarrying of limestone from the Ministry of Minerals, Water and Energy.

Pozzolanas

Deposits of natural volcanic pozzolanas are concentrated in two areas: the Rungwe volcanics near Mbeya in the South West, and the northern volcanics in the Kilimanjaro-Arusha region, which incorporate Mounts Kilimanjaro and Meru and the still active Lengai volcano (see Map 2). The western foothills of Mt. Meru in particular are thickly covered in fine volcanic ash, interbedded with layers of pumice. The latter pozzolanas have been extensively tested in the past and proven to be very reactive (Allen, 1981; Cappelen, 1978); they gave rise to the establishment by SIDO and ITDG of a pilot production plant of lime-pozzolana at Oldonyo Sambu, in the mid 1970s, which for a while produced an acceptable lime-pozzolana binder and blocks. It is thought that the reserves of volcanic pozzolanas are large, but there are no detailed estimates of quantities. Quite a lot is known about the qualities of Tanzanian pozzolanas as well; the Mount Meru deposits were extensively studied by Allen and the BRU (Allen, 1981; Cappelen, 1978), and are considered high quality, whereas deposits near Mbeya were surveyed by SIDO and ITDG, and to a lesser extent the BRU (Sakula, 1980a; Cappelen, 1978; Sakula, 1980b), with good or acceptable results for crushed pumice from Uyole, another sample from Izumbwe, Songwe tuff and ash West of Mbalizi. Reference is also made to the lower quality of the Engaruka deposits in Monduli district and of Tsamasi in Hanang district (Sakula, 1980a).

Volcanic pozzolanas in other countries are known to differ quite a lot in composition and quality, and deposits may not be homogeneous; a positive aspect of most of the Tanzanian pozzolanas is that they are relatively recent deposits, which usually means that the pozzolanas have not lost too much of their activity by weathering. Before they are used more extensively, though, they should be adequately

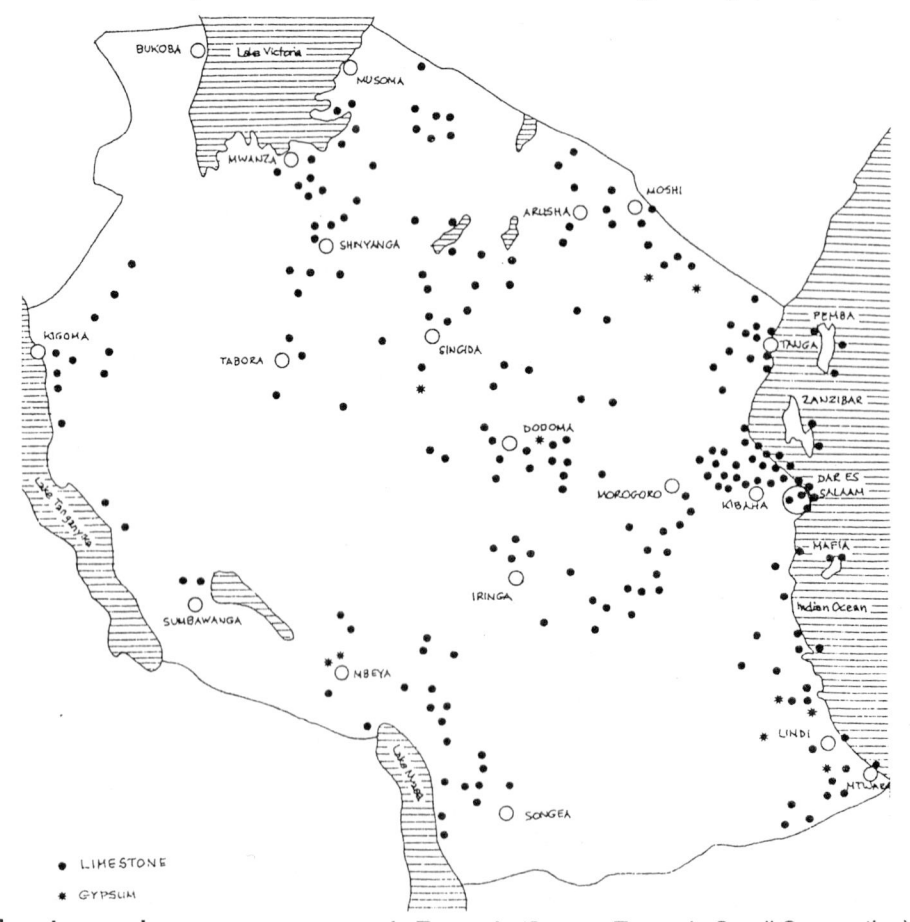

Map 1: Limestone and gypsum occurrences in Tanzania (Source: Tanzania Saruji Corporation)

tested; the most reliable tests for pozzolanas to be used in binders are strength tests on mortar prisms.

There is no history of the use of artificial pozzolanas in binder production in Tanzania. But there are potential sources of artificial pozzolanas which could be further explored, the major ones being rice husk ash, bagasse ash and fired clay, which have been amply proven elsewhere in the world, but it could include less well-known plant ash, fly ash and other materials, which probably will need quite a lot of initial research.

Rice husk ash is obtained by burning rice husks at a temperature of 600-700°C. Rice produces about 20% husks, which make about 20% ash. One tonne of rice could therefore in theory produce about 40 kg of rice husk ash, which could be used with about a third to half of that quantity in lime or cement, to make a satisfactory binder. Rice is grown quite extensively along the coast, in the centre of the country and in the west (see Map 2), but to obtain rice husk ash on a meaningful scale requires working with large rice producers, for instance rice mills. With an annual rice production fluctuating between approximately 600,000 and 900,000 tonnes per year (depending on weather, etc.) between 24,000 and 36,000 tonnes of ash could be produced each year, but it is unlikely that this total amount will ever be available for the production of binders. The SIDO-ITDG lime-pozzolana project tested rice husk ash from Mbeya and Nzega in 1979/1980, and concluded that it would be technically feasible to make a cement out of these materials (Sakula, 1980a).

Waste from sugar cane, in the form of bagasse ash or sugar leaf ash, is another potential source of pozzolana, untried in Tanzania, but used on a large scale in for instance Cuba (Martirena, 1994).

Another artificial pozzolana is clay fired at 600-700°C, which could include underfired products from the brick or tile industry, which is mainly active in the centre, south and west of the country; most of this industry, however, operates on a rather small, decentralized scale, and it would require a large producer or a concentration of small producers to produce sufficient pozzolanic material. In early 1980, the lime-pozzolana project tested a binder made from ground Iringa brick and lime, with poor results (Sakula, 1980a), but one cannot judge the potential of lime-ground fired clay mixtures on the basis of that one test alone.

In comparison to the volcanic pozzolanas, the artificial ones would possibly be available in small quantities; but the volcanic pozzolanas only occur in two major regions of the country, whereas there

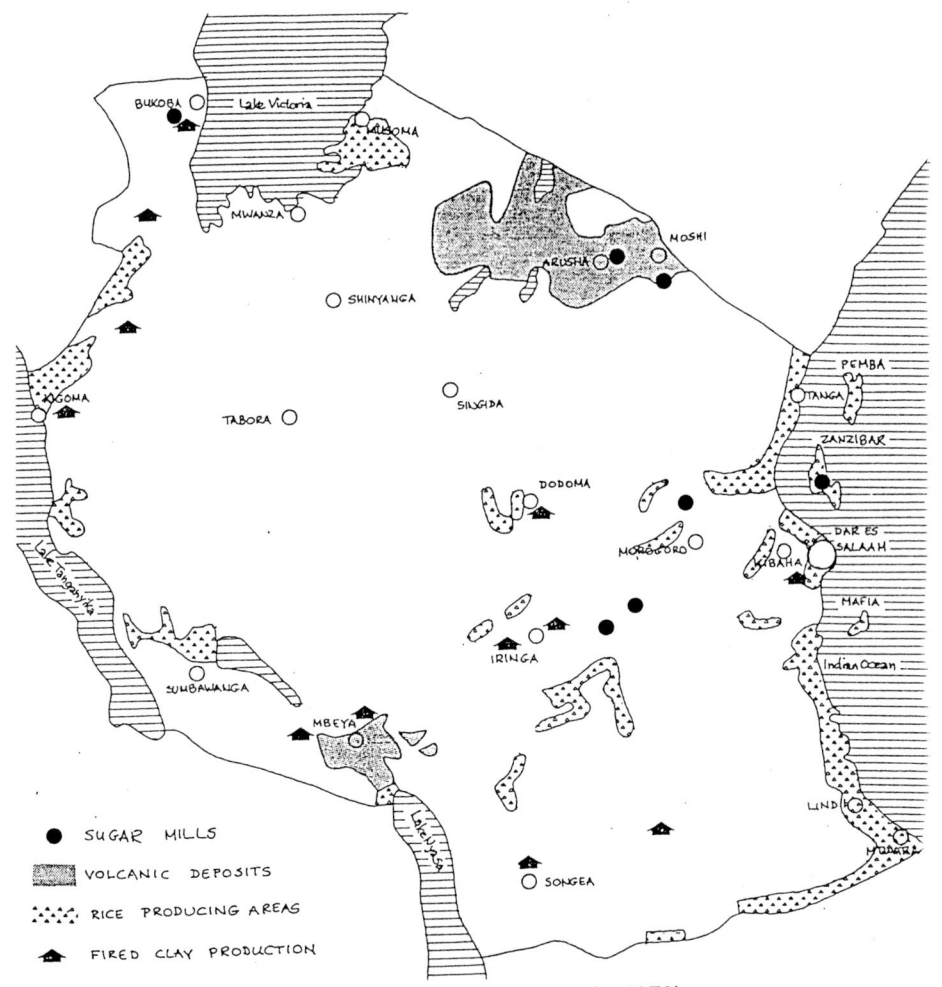

Map 2: Potential pozzolanas in Tanzania (Source: Atlas of Tanzania, 1976)

Tanzania
SECTION 1

is potential for the production of artificial pozzolanas in many more regions.

Gypsum

The cement industry generally uses around 5% of gypsum as a retarder, interground with the cement clinker, but gypsum can also be processed as a binder on its own. Gypsum deposits in Tanzania are shown on Map 1. The ones at Mandawa along the southern coast and at Makanya and Mkomazi in Same district are the main ones which are widely exploited. The authors have no information about the extent or the quality of these deposits. A private company, Timber and Furniture Stores of Moshi, is also burning gypsum from the Makanya deposits, at the moment mainly for the production of chalks, but they have plans to use the material for building panels as well.

Clay minerals

The cement industry uses up to about 0.32 tonnes of clay minerals for making one tonne of Portland cement (Stewart and Muhegi, 1989), but this depends on the purity of the limestone used; these clay minerals are usually available within the vicinity of the cement plant, in the forms of for instance clays, silt or shale. The Mbeya plant uses a local sandstone (Nyiti, undated), the one at Tanga a local red soil, at a rate of about 0.125 tonne per tonne of cement produced *(Daily News, 1982)*.

Fuels

The cement industry typically uses coal or heavy fuel oil for the production of cement clinker. Tanzania's coal deposits being limited to the south west of the country, it is only the Mbeya plant which uses coal as fuel. There have been problems, however, with the supply and the quality of the coal from the Kiwira coal mines, 100 km away, and the plant has had to import coal from Zambia to make up the shortfall; in 1990/91 this amounted to 16,987 tonnes of coal, to a value of 458.6 million TShillings, by far the biggest import component of the company (Nyiti, undated); alternatively, it also uses heavy fuel oil (Rwoga, 1993); the use of local coal as fuel could, nevertheless, save 30% on the fuel bill compared to imported fuel oil. The coal of the south west is also used for lime production in that region.

The cement plants along the coast, in Tanga and Dar es Salaam, currently use fuel-oil, which is an imported commodity, with supply difficulties reported from Tanga due to logistical problems (Nyiti, undated). The Tanga plant is reported to require about 0.1 metric tonne of fuel oil per tonne of cement produced; if both plants were to produce at full capacity, they would require 102,000 tonnes of fuel oil per year. The cost of this oil amounts to about 45% of the operating costs of these plants (Rwoga, 1993).

The larger lime plants generally also use heavy fuel oil; Austroplan (1990) reports a use of 130 litres of diesel oil per tonne of saleable lime hydrate at Tanga Lime, which makes up 44% of the sales price and probably more than half of the production cost of that lime. Other lime kilns using or designed to use fuel oil include a Japanese kiln and a privately owned kiln at Kigamboni near Dar es Salaam, Dar Lime to the North of that town, a kiln at Amboni near Tanga, a rotary kiln at the Southern Paper mills near Mbeya, and privately owned kilns at Monduli, Vitono in Iringa, Kiliwa Mswaki in Hai district, Mbagala and Zanzibar. Since the liberalization of the economy, there do not seem to be major problems in the provision of this fuel.

The smaller lime kilns generally use wood or wood-based (e.g charcoal) fuels. The most primitive type of lime production involves the burning of a heap of limestone on top of a pile of firewood; this is extensively practised along the coast and on the islands, and is very wasteful of fuel: it is estimated that the fuel efficiency of these types of kilns is only about 10%. On the islands, coconut palms constitute the main firewood; they are sometimes supplemented by coconut husks, and smaller size wood. One ITDG report (Bullard, 1988) mentions an average use of 6 coconut trees and 4 oxcarts of smaller wood per heap kiln in Zanzibar; if all 200 estimated heap kilns in Zanzibar were to use this type and amount of fuel, that would be equivalent to 30,000 palm trees per year plus 20,000 oxcarts full of smaller wood. Along the coast, mangrove is also used, for instance in Bagamoyo, Lindi and Mtwara, with grave concerns for the survival of mangrove forests in some locations (Ministry of Tourism, Natural Resources and Environment, 1991). The intermittent kilns, such as the ones established by SIDO in about 30 locations, generally also use firewood, but they are more efficient than the heap kilns; one report mentions the use of 2 m^3 of firewood per tonne of lime produced (Sakula, 1980a).

Overall, lime production in Tanzania may well use in the order of 100,000 trees, or their equivalent, per year as fuel, and some of the uncontrolled cutting which takes place concerns environmentalists. The slightly bigger kiln at Dunga, in Zanzibar, uses a mixture of charcoal, brought from the mainland, and coconut husks; it is unknown whether charcoal is used by kilns elsewhere, though it is known to be a good fuel.

Wood is an ideal fuel for lime burning, because it has a long flame, that penetrates well into the stone; it also does not lead to overburning of the stone, which is a risk with certain fuels that easily reach higher temperatures. In some areas of Tanzania, the supply of wood for lime burning is not yet a problem; these include areas with an original wood cover, particularly to the west, and some areas with plantations; but in other areas wood is much scarcer,

Table 1: Production and performance of the Tanzanian cement industry, 1966-93

Year	TPCC Inst. Cap.	TPCC Ann. Prod.	TPCC Cap. Util.	TCC Inst. Cap.	TCC Ann. Prod.	TCC Cap. Util.	MCC Inst. Cap.	MCC Ann. Prod.	MCC Cap. Util.	TOTAL Inst. Cap.	TOTAL Ann. Prod.	TOTAL Cap. Util.
1966	110	50	46	---			---			110	50	46
1967	110	147	134	---			---			110	147	134
1968	110	157	143	---			---			110	157	143
1969	110	170	154	---			---			110	170	154
1970	110	167	152	---			---			110	167	152
1971	110	178	161	---			---			110	178	161
1972	268	237	89	---			---			268	237	89
1973	268	314	118	---			---			268	314	118
1974	268	296	111	---			---			268	296	111
1975	268	266	99	---			---			268	266	99
1976	268	244	91	---			---			268	244	91
1977	268	245	92	---			---			268	245	92
1978	268	251	94	---			---			268	251	94
1979	520	299	58	---			---			520	299	58
1980	520	286	55	500	20	4	---			1020	306	30
1981	520	253	49	500	137	27	---			1020	390	38
1982	520	219	42	500	155	31	---			1020	373	37
1983	520	131	25	500	142	28	250	13	5	1270	286	22
1984	520	166	32	500	145	29	250	52	21	1270	363	29
1985	520	153	29	500	163	33	250	42	17	1270	357	29
1986	520	229	44	500	162	32	250	47	19	1270	437	34
1987	520	282	54	500	159	32	250	52	21	1270	492	49
1988	520	361	69	500	171	34	250	62	25	1270	594	47
1989	520	320	62	500	190	38	250	85	34	1270	595	47
1990	520	378	73	500	214	43	250	93	37	1270	685	54
1991	520	422	81	500	182	36	250	38	15	1270	641	50
1992	520	378	73	500	228	46	250	75	30	1270	681	54
1993	520	387	79	500	280	56	250	79	32	1270	747	59

and therefore not a fuel to be recommended. However, wood is a renewable fuel and land is not a major constraint in most of Tanzania. Plantations could, therefore, be established and maintained as a source of fuel for lime burning, and a source of secondary income for many rural inhabitants.

Production and consumption figures

Portland cement

Portland cement production in Tanzania is a monopoly, exercised by the Tanzania Saruji Corporation (TSC) and its subsidiary companies TPCC, TCC and MCC. The TPCC started producing at Dar es Salaam in 1966, with a then rated capacity of 110,000 tonnes per year; phase 2, with a rated capacity of 157,500 tonnes per year came into operation in 1972, and phase 3 added another 250,000 tonnes of capacity per year during 1979, bringing the total rated capacity at Dar es Salaam to 520,000 tonnes. In December 1980, the TCC started its operations, with a rated capacity of 500,000 tonnes per year, and the MCC followed in September 1983, with a rated capacity of 250,000 tonnes per year, bringing the total capacity of the TSC to 1,270,000 tonnes per year (Stewart and Muhegi, 1989). Table 1 compiles the actual production figures of the Tanzanian cement industry.

Capacity utilization was good initially, but started to drop in the late 1970s to reach an absolute low of around 22% in the mid 1980s; since then the TPCC has been overhauled by Scancem, who became a minority partner in the company, which substantially increased its capacity utilization, and the country's overall performance. The reasons for the relatively poor performance of the industry are detailed in elsewhere.

Whereas we can find relatively elaborate data on cement production in the available literature, less is known about consumption. Most authors agree that cement has often been in very short supply, and particularly so between the mid-1970s and mid-1980s. In that period, imports often had to make up for inadequate local production, whereas recently there has been a tendency towards more exports. Unfortunately, few detailed figures are available. It is not known whether Tanzania still imports cement, but with a substantially increased local production, and rising exports from that, it is not likely that a lot of OPC is still imported; there may, however, be minor imports of special cements for specific purposes, such as white cement or rapid hardening cement.

The amount of cement available for local consumption nowadays is more affected by exports, but from the available information these appear to have reached a peak around 1990, and subsequently decrease again, which may well be a lasting tendency. Rwoga (1993) estimates that the price increases following price liberalization in mid 1991 have made Tanzanian cement more and more difficult to sell in the export market, and may even start to encourage imports. For an industry which relies heavily on foreign exchange to purchase a number of essential inputs, such as spares, equipment, fuel and expertise, and which has been plagued by bottlenecks in these supplies in the past, it is of course important to be able to export, and to become less dependent on the external assistance, which is anyhow decreasing. But it still leaves the internal markets short of binders.

The per capita cement consumption in Tanzania was on average only 17.2 kg in 1978, with quite a wide variety across the regions (Bengtsson, 1980 and see Table 7). This figure has increased to 22.3 kg per capita in 1988, which is an average growth of 0.5 kg per capita per year, or less than 3%. One UNIDO source (1992) mentions a per capita consumption of about 25 kg per year, presumably for 1990/91, which it considers very low. In reality, figures were probably even lower, though continuing to grow. Taking into account annual distribution figures (Table 6), and assuming a population growth rate of 2.8%, per capita cement consumption for 1990, 1991 and 1992 was respectively 21.6, 22.5 and 22.8 kg/year. It peaked in 1992, to subsequently drop again to 20.8 kg in 1993 and 19.7 kg in 1994.

It is clear that, notwithstanding the establishment of two new cement factories, increasing local production from 245,000 tonnes in 1977 to 747,000 tonnes in 1993, the amount of cement available per capita in the country has increased by no more than a third over those 15 years. Two main factors have contributed to that: a high population growth rate, of about 2.8% per annum, and a shift in tendency from importing cement to exporting it.

Lime

It is very difficult to obtain exact data about lime production and consumption, because few people have looked at the industry as a whole, and a substantial part of the activity takes place in the informal sector, and goes largely unreported.

Unlike cement, which comes as a uniform product with a uniform use in construction, lime is marketed as a variety of products, for a wide range of uses. Limestone lumps are used, particularly on the islands, for ragstone masonry. Crushed limestone is used as an aggregate for road construction and building, ground limestone is used in industry, e.g. in glass production, as fillers and in feedstock, and could be used as agricultural lime; quicklime is mainly used for road stabilization; and hydrated lime is used for construction purposes, sugar refining, mineral processing and other industrial processes.

Ground limestone is produced in a few places only, and largely by the direct users. The glass company, for instance, has its own quarry and was reported to produce about 2,400 tonnes per year around 1988 (Austroplan, 1990). Other producers include Dar Lime, at about 300 tonne per year, according to the same source, whereas the cement

Table 2: Lime manufacturing companies in Tanzania (1982)

Company Location	Shaft kiln	Firewood tonnes	Fuel oil '000 tonnes	Installed capacity tonnes per day	Annual capacity tonnes	Product Price Tsh per bag	Investment Cost Tsh	No of Employees
Bulombora, Kigoma	1	234	-	5	1500	35	80000	9
Magu, Mwanza	1	281	-	6	1800	45	80000	15
Vitono, Iringa	1	168	-	10	3000	45	150000	10
Makere, Kigoma	1	796	-	17	5100	45	120000	10
Ikengeza, Iringa	1	140	-	3	900	35	40000	6
Chemchem, Coast	1	240	-	5	1500	40	120000	12
Hombolo, Dodoma	1	230	-	5	1500	45	50000	10
Simbo, Kigoma	1	140	-	3	900	35	70000	10
Msimbati, Mtwara	1	140	-	3	900	40	75000	10
Magano, Mtwara	1	445	-	3	900	40	75000	10
Kasulu, Kigoma	1	240	-	5	1500	35	60000	16
Tanga Lime	2	-	12	45	4580	38	1.5 million	72
Dar es Salaam (Kunduchi)	2	-	15	50	15000	35	1.5 million	45
	15	3045	27	160	39080	513	3920000	235

industry also has the capacity to sell at least crushed limestone, as it has done for road construction e.g. in Tanga, at 5,000 tonnes per year around 1988, or eventually ground limestone. And there are reports from Zanzibar that substantial, but unquantified amounts of limestone blocks and crushed limestone are used in residential and road construction. Limestone quarrying, be it for cement, lime or dressed stone (marble etc.) production, leads to substantial amounts of quarry waste, which cannot be used in the primary production process; in lime production, the quantities of waste are often between 50 and 80% of all stone quarried. Finding a secondary use for such waste is very important, both in economic terms to make an industry more viable, and in environmental terms; at the moment, this issue has insufficient attention in the Tanzanian binder industry.

Quicklime or hydrated lime can be produced by any type of lime kiln, without or with subsequent slaking respectively. There are three types of lime kilns in Tanzania: large continuous kilns, with a capacity of more than 10 t/d, which are usually oil fired; medium-size semi-continuous or intermittent kilns, typically producing 20-40 t/month, and mainly wood fired; and small heap kilns, of about 3 tonnes capacity, and being wood fired as well. Production figures in all these categories are incomplete, and in some cases no more than estimates. The producers mentioned in Table 2 were reported to be active in 1982 (Kimambo, 1988). Currently, there are said to be considerably more lime producers. There is a total reported installed capacity of large kilns of 222 t/d, with a further 110 t/d in the design or construction phase. If the existing kilns operated for 300 days per year, that would give a total current output of 66,000 tonnes of slaked lime per year; in reality, however, their current joint annual production is probably between 13,000 and 15,000 tonnes per year. This is quite in line with the annual production figure of 18,300 tonnes reported for 1983, which as well as six large kilns included the production of nine village kilns (UNCHS, 1991). With the three planned kilns entering in production as well, the large kilns on their own could produce 100,000 tonnes of hydrated lime per year. In particular:

- Tanga Lime, with an installed capacity of 30-45 t/d (depending on the source: Austroplan, 1990; Ikomba, 1994a), and considered the main producer of lime in 1990, with 3,000-3,500 t/y (Austroplan, 1990), has been out of operation since one of the owners died in 1993. It was being refurbished in early 1994 and expected to start production again;
- Amboni Super Lime, near Tanga, with a similar installed capacity (40 t/d), seems to have nicely filled the gap in the market left by Tanga Lime, and is assumed by one expert to produce about 9,000 t/y (Ikomba, 1994a);
- Kasudeco Lime factory near Kigoma, with an installed capacity of 42 t/d (Austroplan, 1990) is reported not producing at the moment (Ikomba, 1994a);
- the Southern Paper Mills at Mgololo near Mbeya operate a rotary kiln of 60 t/d capacity, to produce about 4,000 tonnes of quicklime per year for its own paper production (Austroplan, 1990);
- a privately owned kiln at Vitono in Iringa, with a design capacity of 15 t/d, but unknown production figures;
- Kigamboni Lime Works in Dar es Salaam, designed for 20 t/d (Ikomba 1994b), but never having attained that level, and currently not producing, for a variety of reasons;
- the Mji Mwema kiln at Kigamboni, built to produce quicklime for the construction of the Dar es Salaam-Lindi road, and reportedly having produced 150 t/month during four months in 1989 (Austroplan, 1990). It has not been in operation for a long time.
- at Monduli, New Arusha Lime has a kiln under construction, of 40 t/d capacity;
- a similar kiln has been designed for Coast Builders in Zanzibar;
- a feasibility study has been done for a 30 t/d kiln for Northern Traders in Hai district.

In the 1970s, SIDO helped to establish 30 village lime kilns (Austroplan, 1990; Ikomba 1994a, 1994c), which were designed to produce lime semi-continuously, but could also be used in batch production. In addition, there are a few privately owned lime kilns, operating according to the same principles, including one at Dunga in Zanzibar and Kunduchi Lime near Dar es Salaam. These kilns could produce about 20 tonnes of lime per batch (Ikomba, 1994b), taking one week at least, but most would not function that regularly, and are more likely to produce 20-40 tonnes per month (Austroplan, 1990). The majority of those kilns, however, is no longer operating. The Austroplan report of 1990 mentions seven (at Vitono, Mkulula, Ikengeza, Usolanga, Simbo, Makere and Hombolo) that are operating, with an annual achievable production of 3,050 tonnes. ITDG has further information about two operating kilns near Maweni and one at Dunga (Holmes, 1990). These figures are confirmed by Ikomba (1994b), who mentions that no more than 10 of the original 30 SIDO-supported kilns are currently working in Tanzania, with an annual production capacity of 4,010 tonnes (Ikomba, 1994b); these are located at Hombolo and Mvumi Mission (Dodoma), Maweni Prison and Kiomoni (Tanga), Makere, Kasudeco and Simbo (Kigoma), Vitono,

Kiwera and Ikengeza (Iringa) (Ikomba, 1994c). In addition, there is the kiln at Dunga, bringing the total capacity to around 4,500 t/y. But there are no figures about the actual output of these kilns, compared to this capacity, except for Ikengeza, which is said to attain 20-25 tonnes per week (Ikomba, 1994c); jointly, they may well reach close to that capacity, maybe 3,500-4,000 t/y.

Heap kilns, also called pit kilns (Austroplan, 1990), are the most basic form of lime production, and the most difficult to gauge. Heap burning of lime is a purely informal sector activity, and since this form of production does not require a kiln, it can easily move site; if one adds to that a variance of activity over the year, due to the influence of the rains, it is clear how difficult it becomes to make an accurate estimate of the number of kilns and the quantity of lime they produce. The most detailed report on heap burning, from Zanzibar (Haasnoot, 1993) mentions that heaps differ in size, but estimates the average output of one heap at 3 tonnes. It further counted no less than 56 heap burning sites within a radius of about 6 km of Zanzibar town centre, mostly at Bububu, North of the town, and at Kiembe Samaki and near the airport to the South; at some sites, a single burner would produce 1-2 heaps per month, but at others a group or co-operative would produce 15-20 heaps. A similar co-operative, Wachumico, just South of Dar es Salaam told one of the authors that they produce 1-2 heaps per week, each making 600 small bags, of two spades full of lime, which is possibly around 3 tonnes per heap; this lime is sold as far as Dodoma and Mwanza! And heap burners interviewed near Amboni reported producing 3-4 tonnes per heap. The Zanzibar report estimates the total lime production from heap burning around Zanzibar town at 340 t/month, or 4,080 t/year. But there are many other sites on the island, including those at Matemwe, Paje and Bwejuu. Another source mentions a lime consumption in 1985 of well over 10,000 tonnes for the island, because there is a wide tradition of using lime, sometimes mixed with cement, in construction (Bullard, 1988), and independent experts estimated the need for quicklime at 20,000 tonnes in 1991 (Holmes and Wingate, 1991), which would equal about 27,000 tonnes of hydrated lime. Zanzibar has only one irregularly operating intermittent kiln, at Dunga, and it must therefore be assumed that the bulk of its lime needs are met by heap burning; it has also been reported that the number of burning sites has increased over the past 15-20 years (Haasnoot, 1993).

Given the above figures, it is not inconceivable that there are currently about 200 heap kilns in operation in Zanzibar, producing about 15,000 tonnes of hydrated lime. Similar heap burning is reported from Pemba, Tanga, Bagamoyo, Dar es Salaam, Lindi and Mtwara. One report concerned with mangroves (Ministry of Tourism, Natural Resources and Environment, 1991) mentions 6 heap burning sites by name in the last region, with the one at Chuno being the most active, with 20-30 heap kilns; it is quite possible, therefore, for Mtwara region to have in the order of 100 heap kilns in operation. On the basis of these estimates, one could extrapolate the nation-wide number of heap kilns to be anywhere between 500 and 800 with an estimated output of 37,500 to 60,000 tonnes of hydrated lime per year.

In summary, the following estimates are made for the current production of quicklime and hydrated lime in Tanzania, calculated as tonnes hydrated lime:

Table 3: Current quicklime and hydrated lime production in Tanzania

Category of kilns	Minimum output (t/y)	Maximum output (t/y)
Large continuous kilns	13,000	15,000
Medium intermittent kilns	3,500	4,000
Heap kilns	37,500	60,000
TOTAL	**54,000**	**79,000**

As can be noticed from the table, the heap kilns are currently the predominant production method. It may be assumed that this is the case, because the other two production methods are doing rather badly, with many kilns being out of operation or operating far below capacity, for a variety of reasons, and also because heap burning is so flexible: it can easily be started or stopped, without major inputs or losses, and therefore adjusts very readily to variations in markets. Whether the heap kilns will remain in that position remains to be seen; the liberalization of the economy now makes it easier for the larger kilns to get access to some of the resources, such as fuel, equipment and refractories, that in the past have often been constraints to them. In addition, heap kilns are seen as environmentally damaging, and measures may be under way to discourage their use in future; on the other hand, they market a cheaper product, which will be in their favour.

Figures for the consumption of lime are as scarce as those for its production. The Austroplan report of 1990 bases its data on contacts with 16 major industries, represented in the Dar es Salaam area; it does not include Zanzibar and Pemba, and seems to have underestimated the informal sector, both in terms of production, as well as use of lime, e.g. in informal construction. A feasibility study, of 1994, by an independent expert (Ikomba, 1994a) is no more complete in its coverage of the industry, but takes more account of the informal sector, although most likely excluding the islands again.

There are no adequate figures at all for the use of

ragstone or crushed limestone for construction or road works. Austroplan estimated the total consumption of ground limestone for 1989 at 10,371 t/y; in reality this may have excluded or underestimated its use for agricultural lime or animal feed. It is likely that the current demand for ground limestone exceeds 20,000 t/y.

The main use of quicklime is currently in paper manufacturing, where the Southern Paper Mills has its own kiln to produce about 4,000 t/y. There is also some use in road construction, estimated at 800 t/y in 1989 by Austroplan (1990), but demand is deemed to be considerably higher than that, at 2,700 t/y in 1992 by Ikomba; the latter figure is based on the construction of 135 km of new roads per year, at 20 tonnes of quicklime per km, and an assumption of no use of lime in road maintenance (Ikomba, 1994a). The current consumption figure of quicklime may therefore be estimated at about 7,000 t/y, which is equivalent to about 9,300 t/y of hydrated lime.

Users of hydrated lime, outside construction, include the sugar industry, using 3,400 tonnes in 1989 (Austroplan, 1990), and an estimated 3,800 tonnes in 1992 (Ikomba, 1994a); adding the needs in Zanzibar (Holmes, 1990), current consumption could be 4,000 t/y. The paint industry used about 550 tonnes in 1989 (Austroplan, 1990), and mineral processing about 720 tonnes (Austroplan, 1990), whereas tanneries account for 675 tonnes in 1992 (Ikomba, 1994a). About one third should be added for unreported and minor users, bringing the total current annual consumption of hydrated lime, outside construction, to about 8,000 tonnes.

The great unknown factor is the use of hydrated lime in construction; a lot of building, particularly of shelter, takes place in the informal sector, and goes unreported; it becomes even more difficult to trace such consumption, if the lime used is also produced by the informal sector, in heap burning. There is a large tradition of building with lime on the islands of Zanzibar and Pemba, and to some extent along the coast. Traditionally, only lime was used in building shelter, nowadays, it is often mixed with cement. One source reports an average use of 8 tonnes of lime for the construction of a modern 60 m house in Zanzibar, whereas the conservation of old houses in the stone town might require as much as 25 tonnes per house (Bullard, 1988). The estimated production of 15,000 tonnes of lime per year in Zanzibar, mentioned above, will be largely used in construction, with some also going to roads and the sugar industry. It is quite conceivable, that the total annual use of lime in building on the islands amounts to at least 20,000 tonnes. Ikomba further estimates a consumption of another 20,000 tonnes of lime for building on the mainland, in 1992 (Ikomba 1994a), which is probably a conservative figure compared to the islands, but the mainland has less of a lime tradition. The total consumption of lime in building in Tanzania is therefore estimated at around 40,000 tonnes per year at the moment; it should be noted that this is a very rough figure, which should be subject to further verification.

In summary, the estimated annual consumption of quicklime and hydrated lime is currently as shown in Table 4 (converted to tonnes of slaked lime where applicable):

Table 4: Estimated annual consumption of quicklime and hydrated lime in Tanzania (t/y)

Industrial and road use of quicklime	9,300
Industrial uses of hydrated lime	8,000
Use of hydrated lime in building (approx)	40,000
TOTAL (approx)	**57,300**

The market of lime can also be influenced by imports and exports; there is no known evidence of the latter. There have been substantial imports of lime reported for the 1980s (Ikomba, 1994a), reaching a peak of 7,010 tonnes in 1985. In those years, the chemical and paint industries imported 1,000-3,000 tonnes each year, with the remainder going to the construction of Makambako-Songea road. But the same source mentions there have been no lime imports during the last five years. This would put the current annual consumption of lime at 57,300 tonnes, which is well within the production range mentioned previously.

Pozzolanic binders

Up till very recently, the Tanzanian cement industry did not make use of any pozzolanas in cement production, but this has now changed with the MCC going into the production of a Portland-Pozzolana Cement, using its own pozzolana quarries near Mbeya. At this moment, the authors have no detailed information about the composition and the characteristics of this cement, nor about the quantities produced.

There has been one attempt to produce a lime-pozzolana, at Oldonyo Sambu, on the Western flanks of Mount Meru. This was in the late 1970s, in a joint SIDO-ITDG project. This experimental plant had an intermittent lime kiln of the type established by SIDO all over Tanzania, and used the very fine pozzolanas of the area; it produced a binder as well as prefabricated building blocks, using this binder and pumice as aggregate. The mixture used for the binder made at Oldonyo Sambu had equal weight proportions of lime and pozzolanas. With a lime kiln of a capacity of 3 tonnes of lime per day, this plant could therefore in theory produce 6 tonnes of binder per day, or about 1,800 t/y, but it was deemed that a more realistic output would be 1 tonne of lime per day, or 600 t/y of binder (Holmes and Wingate, 1991). The plant was only ever used, however, for two periods of trial production, in 1978 and 1979, and about 30 tonnes of lime, 10 tonnes of lime-pozzolana and 5,000 lime-pozzolana blocks were produced (UNCHS, 1985). The plant was handed over to the village in the second half of 1980, but ceased production shortly after. Various attempts to revive it with other producers have not succeeded.

Gypsum

There is one producer of gypsum in Tanzania, the Timber and Furniture Stores of Moshi, who started to produce gypsum in small quantities in 1993. They use a small vertical kiln built in brick masonry, which is fired for 3.5-4 hours per batch, using charcoal as fuel; each batch produces 190 kg of gypsum. The kiln could therefore produce up to 114 tonnes of gypsum per year, but it is unknown whether this figure is actually reached. The product is currently mainly used to manufacture chalks for schools, but the industry has an interest in gypsum as a building material and for the production of prefabricated panels, for which it has some potential.

Summary

Table 5 below summarizes the production and consumption data dealt with in detail in this chapter, using the confirmed figures for the cement industry for 1992, and estimates for the lime industry. Figures are rounded to the nearest thousand, and in the case of lime they are given for the hydrated lime equivalent. It is clear from the table that cement is the predominant binder in Tanzania, with over 90% of the market. There is substantial production of lime as well, particularly in heap kilns. Gypsum and pozzolanas are negligible for the moment. It is to be noted furthermore that capacity utilization, although it has improved in the cement industry, is still not satisfactory, and it is also not good in the larger-scale lime production.

Production details

Cement

Tanzania's oldest cement plant, TPCC at Wazo Hill just north of Dar es Salaam, started operating with a single line of 110,000 t/y rated capacity in 1966; a second line of 160,000 t/y capacity was commissioned in 1972, and a third line, of 250,000 t/y capacity commissioned in 1979, bringing the current capacity to 520,000 t/y. The plant uses dry process technology with rotary kilns. The plant was initially under German management, up to 1973, and subsequently under Indian management up to 1981. It was then under local management for several years, until a major overhaul started, with Swedish assistance, in the mid-1980s. The plant is currently managed by a SCANCEM management team.

Table 5: Production and consumption of binders in Tanzania, 1992 ('000 t/y)

Industry	Capacity	Production	Consumption
TPCC	520	378	343
TCC	500	228	201
MCC	250	75	70
Total cement	**1,270**	**681**	**70**
Crushed limestone	?	>5	>5
Ground limestone	?	>20	>20
Total limestone	**?**	**>25**	**>25**
Large continuous lime kilns	67	13-15	?
Medium-size intermittent kilns	5	3-4	?
Heap kilns	>100	38-60	?
Total lime	**>172**	**54-79**	**57**
Pozzolanic binders	2	0	0
Gypsum	0	0	0
OVERALL TOTAL	**>1,444**	**>760**	**>696**

Both SCANCEM and Swedfund have each taken a 13% share in the industry, with the TSC retaining the remaining 74%, but it is planned that more of that share will be sold to SCANCEM and Swedfund, the employees and the public at large (Rwoga, 1993). Until 1979, capacity utilization at the plant was high, at above 90%, but it then started dropping, to reach a low of 25% in 1983; the subsequent overhaul has improved the situation, and the factory is now again attaining around 80% capacity utilization (Stewart and Muhegi, 1989; Nyiti, undated; Rwoga, 1993). The poor performance of the cement industry in general in the mid- to late 1980s has been attributed by the same sources to, amongst others:

- a lack of foreign exchange to buy spares, equipment, fuel etc.
- a gradual rundown of the industry due to lack of maintenance, influenced by the lack of spares and the lack of industrial repairs capacity in the country,
- constraints in the fuel supply, particularly in Mbeya and Tanga,
- frequent power shedding, particularly in 1992 due to the drought,
- lack of quarrying and haulage equipment for raw materials,
- insufficient technical and management experience,
- inadequate motivation of the labour force.

Most of these factors will have contributed to the poor performance at the TPCC. The situation has, however, improved following the factory's rehabilitation, and the company is now doing relatively well; whether it will maintain this performance will depend on preventive maintenance in future. One of the advantages of this factory over the two others is its relatively good access to markets, with the city of Dar es Salaam on its own consuming over half of the factory's output. Another positive element is the increase of exports over recent years, which gives the industry some access to foreign exchange; all factories now operate external accounts, which they can use to import essential inputs, subject to control by the Bank of Tanzania. The TPCC uses heavy fuel oil as energy, which is imported. The fuel oil accounts for about 45% of the total operating costs, with electricity requiring another 15%, and paper bags 10% (Rwoga, 1993). We do not have any current details about the other inputs, the investment and manpower levels.

The factory at Tanga is a single dry-process line, of 1600 t/d capacity, or 500,000 t/y. It came into operation in late 1980, after 4 years of feasibility studies and 6 years of construction. The construction was delayed with more than a year due to adverse soil conditions, supply bottlenecks and damage to equipment ordered from abroad. The commissioning also took much more time than expected due to inadequate quarrying equipment and delays and constraints in supplying fuel oil, diesel oil and explosives, and power cuts. The overall investment, at the 1976 contract rate was 538 million TSh in foreign exchange, mainly provided by a Danish soft loan and a Dutch grant, and 292 million TSh in local money, or a total of 830 million TSh. The investment level as per 30 June 1994 was estimated at 2.6 billion TSh, or $5.2 million, after depreciation. The factory is relatively modern, with sophisticated instrumentation and pollution control; it has a packing capacity of 200 t/h. At its time of construction, provision was made for an expansion to 1 million t/y capacity, but this has not been realized so far *(Daily News, 1982)*. The capacity utilization at Tanga has always been far too low, between 27 and 38% in the 1980s (TSC, Table 1); a range of reasons for this are given in the previous paragraph; it seems that supplies of fuel oil and inadequate quarrying equipment have been major factors; the latter has been recently solved (Nyiti, undated).

A rehabilitation programme for TCC, financed by Danida, is scheduled for 1992-1994, and has started to raise capacity utilization, to 56% in 1993. TCC is wholly owned by the TSC, who is actively looking for joint ventures with foreign partners and has received some offers (Rwoga, 1993). It has been under foreign management from the start. Currently the Swiss firm Holderbank has a management contract. In 1982, the factory was employing 640 people, some of which were trained locally, and some abroad; this included 8 foreign experts and their 8 local counterparts; the factory currently employs 558 people. Assuming 2,000 manhours worked per person per year, it took therefore 8.8 mh/t to produce the 145 tonnes output of 1982, but productivity has now risen, to reach around 4 mh/t in 1993. A similar size industry in India would use about 3 mh/t, and this is achievable in Tanga with yet higher efficiency after full rehabilitation (Sinha, 1990). These figures are still high compared to the more capitalized cement industry in the North, which is typically using between 0.3 and 1.0 mh/t for a factory of this size (Blue Circle Group, 1982).

In 1982, the factory was also reported to have a daily fuel requirement for full production of 160 metric tonnes of fuel oil for its rotary kiln *(Daily News, 1982)* which makes 0.1 tonne of fuel oil per tonne of cement, or approximately 4 MJoule of kiln energy, per kg of clinker. Current fuel consumption for the kiln varies between 3.6 and 4.2 MJ per kg of clinker, which is about right for a dry production line of that size (Hill, 1993). To this, one would have to add the electricity required for grinding the clinker to cement, which typically would add another 0.4 MJ/kg (UNCHS, 1991b), to reach a total energy requirement of 4.0-4.6 MJ/kg cement produced at Tanga. The fuel oil makes up 33% of the production costs, with electricity taking another 16%; energy on its own therefore accounts for about half of the cost of cement. The other items are bags (11%), gypsum (6%), maintenance (15%), consumables (10%) and manpower (10%).

The factory in Mbeya started to produce in 1983, ten years after the inception of the project; it had been installed two years previously, but delays in the power supply put commissioning off. It consists of a single dry process line, of 250,000 t/y capacity. Most of the equipment was supplied from Denmark, including a crusher, water treatment plant, raw mill, cement mill and packing plant; this equipment was said to be in good shape in 1991 (Nyiti, undated). The total investment was estimated, in 1991 figures, at 5,003 million TSh own capital, and 1827 million TSh borrowed capital, making a total of 6,830 TSh. As per 30 June 1994, it was estimated at 3.6 billion TSh, or $7.2 million, after depreciation. The factory is wholly owned by the TSC, who is actively looking for foreign partners to rehabilitate it and assist in the management, but sofar unsuccessfully.

The MCC has been under local management from the much delayed start of production. Lack of experience has certainly contributed to the plant's poor performance over the years, but there have been other factors as well, notably problems with coal supplies, electricity, raw materials haulage and spares supplies. The factory never made a profit, has not been able to pay its creditors, and had accumulated losses, which stood at 1,530 million TSh, or US$6.8 million in 1990 (UNIDO, 1992). Its capacity utilization has always been poor, between 15 and 37% (TSC, Table 1); this is largely due to the above problems, but there may also not have been a sufficiently large natural market in Mbeya and surrounding regions, which one source puts at 80-100,000 t/y (UNIDO, 1992).

Even with exports to Malawi increasing, the MCC would have to find markets further away in Tanzania; the use of coal, which is a cheaper fuel than oil, and the shift to PPC would, however, enable the factory to compete well with the others, even at further distances. The factory needs an overhaul, to rehabilitate existing machinery, buy some new equipment especially for quarrying and restore stocks of spares, and expert assistance for a while (Rwoga, 1993). The current manpower level at the MCC stands at 590, which UNIDO considers unnecessarily high (1992). Indeed, productivity is substantially worse in Mbeya than in Tanga, with 14.9 mh/t of cement produced in 1993, and still 12.7 mh/t in its peak year, 1990. But even so, labour still accounts for only 8% of the production costs. Current fuel consumption per kg of clinker stands at 4.2-5.0 MJ; the MCC uses both fuel oil and local coal. Adding the grinding energy, total energy consumption would be 4.6-5.4 MJ/kg cement. The furnace oil was accounting for 8% of the production costs, coal for 15% and electricity for 20%, making a total of 43% for the energy component of production. Other costs included bags (12%), gypsum (6%), maintenance (4%), consumables (5%) and labour (8%). (Note: these percentages are provided by the TSC. The total falls substantially short of 100% and, in reality, they may have to be adjusted upwards).

Lime

Most of the large oil-fired lime kilns are of the vertical shaft type with oil burners along two sides of the kiln, about one third from the bottom; but there are two noticeable exceptions: the Southern Paper Mills have a rotary kiln, and the Ministry of Works owns a very short shaft kiln at Mji Mwema, fired from the top.

Tanga Lime Works, some 30 km West of Tanga, is one of the earlier vertical shaft kilns (VSK), commissioned in 1974. Its design capacity is variably put at 25 t/d (Austroplan, 1990), 30 t/d (Holmes, 1990), or even 45 t/d (Ikomba, 1994b), whereas the average production is estimated at 15 t/d, or 3,000-3,500 tonne of hydrated lime per year (Austroplan, 1990). The kiln used to operate continuously with large orders, but this is not popular with the staff, and batch firing on a 8-hour cycle, during a 5 day week, was reported to be more common in 1990 (Holmes). This may explain the below capacity output. Following the death of one of the owners, the kiln has not been operating since early 1993. The kiln is built in brick masonry on a concrete foundation, and lined inside with refractory bricks, which are said to last up to 12 months; imported refractories are now amply available. The kiln is charged with fist size lime stone via an inclined skip hoist, and discharged through two openings at the bottom with vibrating chutes. Eight oil burners, four along two sides of the kiln, are fed with heavy fuel oil, propelled into the kiln via air jets; this firing system is requiring a lot of maintenance; the tops of the burners, made locally from galvanized pipe, have to be replaced every 6 hours.

The kiln uses 130 litres of heavy fuel oil per tonne of saleable lime hydrate (Austroplan, 1990), which is equivalent to approximately 5.2 MJ/kg of lime. The same source complains about the burners not evenly distributing the fuel through the kiln, leading to high levels of unburnt limestone (Loss on Ignition of 23.5% in the final product). The burnt limestone is slaked manually on the shed floor, sieved manually, milled by hammer mill, and bagged into 25 kg bags. In 1990, the fuel oil cost amounted to about 44% of the sales price, electricity to 7% and manpower to 5.5% (Austroplan). When in operation more recently, the kiln used to employ 87-90 people (Austroplan, 1990), resulting in a productivity of 50-60 mh/t. Earlier figures were more favourable, with a production of 4,580 tonnes produced by 72 people reported in 1982, which is a productivity of 31.5 mh/t (Kimambo, 1988).

Amboni Super Lime, about 10 km North of Tanga, is owned by Mr. Marshed. It is a similar kiln to the previous one, with a capacity of 40 t/d according to the owner (also Holmes, 1990), and 45 t/d according to Ikomba (1994b). Its refractories are said to last 6 months in the burner zone and 12 months elsewhere. The limestone from the quarry is broken by a roller crusher; the discharge of the kiln is done every four hours approximately. The burnt limestone used for road stabilization is only sieved, the one used for the sugar industry is slaked in a batch hydrator, then milled in a hammer mill. The kiln is said to use about 2,000 litres of fuel oil per day (Holmes, 1990), which would be too low a figure for a 40 t/d output. The same source mentions that about 10% of unburnt lime is produced, which is returned to the kiln. Since Tanga Lime is not producing, Amboni Super Lime is doing well at the moment.

A smaller VSK, of 20 t/d design capacity, is owned by Mr Assenga at Kigamboni, just south of Dar es Salaam. This kiln had not been operating for a while during a visit by one of the authors in early 1994; it seems, there were problems with the firing, leading to a lot of underburnt material, and possibly also with marketing. For more details on this kiln, see the case study in Section 2.

Other VSKs in Tanzania, already existing or in

the design or construction stage, are essentially similar to the ones above. The required investments for such kilns ranges from 10 to 30 million TSh, or $20,000-60,000, depending on the capacity and level of mechanization (Ikomba, 1994a and b), with the kiln requiring about 50-55% of that amount, and machinery and equipment 20-30%.

The kiln at Mji Mwema, in Kigamboni, owned by the Ministry of Works, was built for the Dar es Salaam-Lindi road project. It is described by Austroplan (1990), as being quite mechanied, having a crusher for the limestone feed, skip hoist and conveyors. The kiln itself is a steel structure, with a burning lance firing from the top, fired by diesel oil; it is emptied by a rotary table feeder; the burnt limestone is ground by hammermill and screened, and used as quicklime. The kiln capacity is given as 2-6 t/d of quicklime; during the first four months of production, in 1989, the average output was 150 t/m. The plant used to employ 3 shifts of 4 people, and a further 20 in the quarry, which makes 33 employees, including a manager; and the average productivity would have been 34.6 mh/t. But this kiln has not been in operation for years now.

According to Austroplan (1990), lime production at the Southern Paper Mills uses quite a different technology: a rotary kiln with satellite coolers, of a capacity of 60 t/d; but the actual production in 1988 only stood at 4,000 t/y, which is less than a third of the kiln's rated capacity. The kiln uses an unknown quantity of diesel oil. The plant has no quarry of its own and is said to purchase its limestone from the MCC or Dar es Salaam. During the paper making process, calciumcarbonate is produced as a by-product, which is pelletized and fed back into the kiln with 25% of new limestone.

Most of the village level lime kilns in Tanzania are of the type promoted by SIDO in the mid 1970's, which is the Indian Khadi and Village Industries design; they were meant to be operated semi-continuously, but often used as a batch kiln. In addition, some other intermittent or batch kilns are known to exist in the country, e.g. at Dunga in Zanzibar and at Maweni near Tanga.

SIDO helped to install 30 kilns, of different sizes, with capacities ranging from 1-10 t/d. Their diameter would range from 1.20 to 1.80 metres, and their height from 4.5 to 15 metres; it would cost 1.2 to 2 million TSh, or $2,400-4,000, to install such a production unit (Ikomba, 1994c). In the late 1970s, only 20 of those kilns were reported to operate well, and that number has now dropped to ten (Ikomba, 1994c and Chapter 2); the same source estimates that lack of incentive, profitability and extension work have contributed to that drop.

The small kiln, operated for a while by SIDO and ITDG at Oldonyo Sambu, has a 1.20 m diameter by a height of 4.50 m. It has 1 m thick masonry walls, with steel stirrups. Its maximum output was 3 t/d, but 1 t/d was reported as more achievable (Sakula, 1980a; Spence and Sakula, 1980). This kiln required 3 tonnes of limestone per tonne of lime produced, and 2 m^3 of firewood (Sauni, 1981); if the latter is assumed to weigh about 1 tonne, that would mean an energy consumption of about 15 MJ/kg lime, which is nearly three times the requirement of a large oil-fired VSK. This kiln has not been operative for a long time.

A currently operating larger kiln, at Ikengeza village in Iringa, is producing 20-25 tonnes of lime per week, using ten employees, who are paid on a monthly basis and also retain a share of production (Ikomba, 1994c). This puts their productivity at 16-20 mh/t, which seems very high compared to the larger kilns, and may have not taken account of limestone quarrying. Comparable figures for the VSK at Dunga, Zanzibar, are that 38 labourers produce 50 tonnes of quicklime (equivalent to about 66 tonnes of lime hydrate) in a week, which is a productivity of 23-28 mh/t (Suleiman, 1994).

Holmes (1980) reports on another SIDO-kiln at Maweni being operated in batch rather than continuously, since it fits local customs better. It is also more in line with the local tradition, using hill side batch kilns, of 3 m diameter and 5 m height, to produce about 10 tonnes of lime in a three day cycle.

According to Ikomba (1994c), those village kilns that are doing well seem to have moved from being run by villages as co-operatives, using the villagers' voluntary labour, and producing mainly for own needs or to gain income for the village, towards commercial enterprises, where villagers are employed to work in lime production, where accounts are properly maintained, etc. As in the case of Ikengeza, they can still help a village to realize its building projects, as well as supply a market in the district. For village lime burning to be successful, it needs to be on a commercial basis, and certain skills have to be present. Also, several of the existing kilns and many of the ones no longer producing need to be rehabilitated.

Lime production using heap burning is taking place entirely in the informal sector, and therefore not registered. It has been estimated earlier in this paper, that there may be 500-800 heap burning sites in Tanzania, producing an average of 3 tonnes of lime every two weeks, or in total 37,500-60,000 tonnes per year. These are average figures, though, and very much estimates. Heap burning is, for instance, influenced by the weather, and may be much reduced during rainy periods. Heap burning does not require a huge investment; in one case, of Amboni Quicklime described in Section 2, it amounted to TSh 572,000, about $1,140, with another TSh 183,000, $360, required for the start-up of operations, or a total of $1,500, but this is a relatively large heap burning operation, with a production of

Table 6: Regional distribution of cement in Tanzania, 1978 and 1988

	1978			1988		
REGIONS	POPULATION	CEMENT CONSUMPTION (TONNES)	PER CAPITA CONSUMPTION (KG)	POPULATION	CEMENT CONSUMPTION (TONNES)	PER CAPITA CONSUMPTION (KG)
ARUSHA	928,476	15,477	16.7	1,352,225	17,659	13.1
COAST	516,949	19,695	38.1	639,182	7,405	11.6
DAR ES SALAAM	851,522	105,241	123.6	1,360,850	211,484	155.4
DODOMA	971,921	13,188	13.6	1,235,277	17,162	13.9
IRINGA	922,801	11,996	13.0	1,193,074	14,566	12.2
KAGERA	1,009,379	3,023	3.0	1,313,639	9,179	7.0
KIGOMA	648,950	3,447	5.3	853,263	3,862	4.5
KILIMANJARO	902,394	20,212	22.4	1,106,068	43,095	39.0
LINDI	527,902	4,312	8.2	642,364	3,309	5.2
MARA	723,295	6,446	8.9	952,616	7,569	8.0
MBEYA	1,080,241	14,244	13.2	1,476,261	30,130	20.4
MOROGORO	939,190	20,925	22.3	1,279,931	15,424	12.1
MTWARA	771,726	7,491	9.7	887,583	11,596	13.1
MWANZA	1,443,418	15,530	10.8	1,876,776	16,604	8.9
RUKWA	451,897	1,841	4.1	704,050	3,691	5.2
RUVUMA	564,113	3,033	5.4	779,868	4,708	6.0
SHINYANGA	1,323,482	3,918	3.0	1,763,960	6,555	3.7
SINGIDA	614,030	2,260	3.7	793,887	1,814	2.3
TABORA	818,049	6,549	8.0	1,042,622	7,638	7.3
TANGA	1,038,592	21,114	20.3	1,280,262	51,906	40.5
ZANZIBAR & PEMBA	479,235	1,670	3.5	640,685	31,089	48.5
TOTAL	17,527,562	301,612	17.2	23,174,443	516,444	22.3

Sources: TSC (1994), Komba (1980), Bureau of Statistics (1988)

10 tonnes per week, and at other sites investments are likely to have been even lower.

One source describes heap burning in Zanzibar (Bullard, 1988) as using heaps of varying sizes, producing 6-30 tonnes of lime hydrate per batch; this latter figure seems high compared to figures from other sources. The heap diameter is put at 3.5-8 m, with a fuel height of 0.9-1.2 m, and a coral dome on top of 0.8-1 m.; this coral is in 75-150 mm lumps. The fuel has a small layer of thin poles at the bottom, of 2-3 cm diameter, and is using coconut poles of 15-25 cm diameter and about 1 m. length above that, filled in with husks. Another source on Zanzibar (Haasnoot, 1993) roughly confirms this kiln buildup, and also mentions that the outer layer of uncalcined stone is removed after the firing, and the remainder slaked, which would provide about 3 tonnes of lime per kiln, which is a figure in line with that found along the coast of the mainland. Such a kiln requires on average 22 oxcarts of 0.5 tonnes, or 11 tonnes of limestone.

On average, it takes two people 6 working days each to produce the average 3 tonnes of lime, which amounts to 32 mh/t; this is likely to exclude quarrying but to include obtaining fuel. Another source mentions that quarrying of 1 tonne of limestone may take 1-2 mandays; this would amount to 29-58 mh/t of lime (Suleiman, 1994). This would raise the total labour requirement per tonne of heap burned lime to around 60-90 mh. In a few cases, though, quarrying is made easier by using explosives (Reuben, 1994); in one such case, 12-14 workers produced 10 t of lime per week, at a productivity of 48-56 mh/t; this included a substantial amount of time for transporting water to the production site. A small heap kiln, producing 3 tonnes of lime, would require approximately 13 m^3 of densely packed fuel, including coconut logs, smaller wood and coconut husks, estimated to weigh about 7 tonnes; this makes around 35 MJ/kg lime produced, which is a very low fuel efficiency. Holmes (1991) estimates that Zanzibar might require 600-800 ha of fuelwood plantation, to produce the lime it needs.

This type of production in the coastal belt of the mainland is largely similar, though in some places mangrove is used as source of fuel. Both the Wachumico Co-operative, at Kinondoni just South of Dar es Salaam, and heap burners at Amboni seem to produce around 3 tonnes of hydrated lime per heap, which is sold in plastic bags.

Pozzolanic binders

One pozzolanic binder that has so far been produced is a lime-pozzolana, made in the late 1970s at Oldonyo Sambu. The lime was produced in a semi-continuous lime kiln, of the SIDO type, of a rated capacity of 3 tonnes per day, described in the previous section. It used about 2 m^3 of firewood per tonne of lime produced. The pozzolanas were dug locally, and were so fine, that they did not need any further treatment. Lime and pozzolana were blended in equal parts, using shovels, then sieved on a 3 mm sieve to remove any slight lumps, before being packed in 20 kg bags. The plant was originally a pilot plant, owned by SIDO, but handed over to the village after pilot production. Although technically quite successful, the product failed in the end, because there was not enough of a local market, it lacked transport, and it lacked management capacity. It might have succeeded, had it been established by a small entrepreneur and with the market of Arusha in mind.

Recently, the MCC has started to produce a Portland Pozzolana cement, using its existing plant. Since the volcanic pozzolanas used do not need to be fired, they can be added at the grinding stage, which should normally increase production and productivity levels, and reduce the amount of energy used per kilo of cement. But there are no details available about this as yet.

Gypsum

The one small production unit of gypsum at Moshi uses a jaw crusher to reduce the dimensions of the rock to about 2 cm; this material is subsequently washed and sundried. A hammer mill then further reduces the raw material, which is subsequently calcined in a small vertical kiln at a temperature of 160°C, fired with charcoal. The material enters the kiln from the top; a vertical axis with paddles is stirred by hand, to guarantee good calcination. The kiln is fired twice daily, for 3.5-4 hours, and produces 190 kg of gypsum per batch.

Use and distribution of binders

Cement

Cement is mainly used for building, and for some infrastructural works. Since these works take place all over the country, cement use is also widespread. Due to differences in function, wealth, etc., it is used in much greater quantity in urban than in rural areas. But its use is also affected by affordability, which is both influenced by relative income levels and wealth in general, and the costs of cement, which may vary from region to region due to transport costs. Availability, influenced by transport capacities, etc., is another important factor explaining regional differences. Finally, some regions are far more populated than others, and therefore require more cement. Local habits and skills, as well as the availability of local alternatives, can influence regional demand for the material. All these factors, and possibly others, help to explain the differences in the regional distribution of cement, in tonnes, as given in Table 6 (Muhegi, 1983 and TSC, 1994). Unfortunately, we do not have details of the amounts of cement used for different building purposes, e.g. housing, other buildings, infrastructure.

We also have figures of the regional population distribution, of the census year 1978 (Komba, 1980), and of the census year 1988 (Bureau of Statistics) which, combined with the cement distribution figures, give us an idea of the regional per capita consumption of cement, as presented in Table 7. The average per capita consumption of cement in 1978, of 17.2 kg or about one third of a bag, was low, even for a developing country. This has steadily risen to 22.3 kg per capita ten years on, but has not substantially increased since then. The table also shows a wide regional disparity, with the heavily urbanised Dar es Salaam region consuming about 7 times the national average, that is 123.6 kg per capita in 1978 and 155.4 kg per capita in 1988, and regions faced with transport constraints, such as Kagera, Kigoma, Rukwa, Ruvuma, Shinyanga and Singida in the range of 3 to 5.4 kg per capita in 1978, and between 2.3 and 7.0 kg in 1988.

The establishment of two new cement factories, in Tanga in 1980 and in Mbeya in 1983, has clearly had some impact. Cement consumption in Tanga region has noticeably increased, and so did it in neighbouring Kilimanjaro region; and consumption in Zanzibar, now supplied by boat from Tanga, has shot up dramatically. Consumption has also gone up in Mbeya region, but not noticeably in surrounding regions. Establishment of these new factories did not really improve supplies to more outlying areas.

Cement distribution was fairly controlled and regularized at the start of the above period. What is more, up to 1982, a freight equalization charge was applied to share the costs of transport equally over the country, but this lead to a high average price of cement. Distribution measures, via regional and district trading companies, and the freight equalization, favoured cement supplies to outlying areas, but, in reality, sales were probably hampered by transport constraints; it, however, made cement consumed close to the factories relatively expensive. With the abolition of the freight equalization charge in 1982, cement prices began to vary throughout the country, and this will undoubtedly have affected regional demand. In 1982, for instance, the retail price of a tonne of cement stood at $111; on top of that, the railways would charge up to $40 to transport a tonne to some of the outlying regions such as Kigoma or Kagera, and about the same rate would apply for transport by boat from Tanga to Mtwara or Lindi (Muhegi, 1983). Unfortunately, Tanzania Railways only had the capacity to transport about 20% of the requirement in 1981 (Bengtsson, 1980), so much of the cement had to be transported via the much more expensive road transport; this would cost about twice as much on good tarmac roads, but in the case of transport on badly maintained earth roads, such as to Kigoma, the price of transport would be prohibitive, and could reach $288 per tonne (Stewart and Muhegi, 1989). Thus, at that time, a bag of cement in Kigoma could cost $20, which was the equivalent of 40% of a monthly minimum wage.

It is clear then that cement has been a luxury material in outlying areas, particularly the North West of the country, and the more so since 1982. Currently, ex-factory cement prices stand at $100 per tonne at Tanga and Mbeya, and at $100-110 in Dar es Salaam, but these prices can double or triple once the cement arrives in outlying regions. The TSC currently charges $0.15 per tonne/km for the transport of cement by road; for Kigoma, this would be approximately $230 per tonne (it would cost $100 per tonne by rail); once profits and retail costs are added, cement may still cost $20 per bag to a user in Kigoma.

But apart from the railways, road transport was also insufficient in the early 1980's. In 1980, the TSC estimated that it needed 136 trucks for road transport, 350 general cargo rail wagons and 25 bulk rail wagons, one ship plus various other facilities, to adequately handle and transport cement all over the country (Muhegi, 1983), but they only expected to obtain all bulk wagons, the ship and five trucks within the next year. In early 1981, the Saruji Trucking Company was established, in order to establish an internal transport capacity for the industry, but the size of that company has always remained small (Rwoga, 1993). According to Muhegi (1983), the Tanga and Dar es Salaam plants together only possessed 13 trucks and 7 trailers in 1983, with a total capacity of 321 tonnes; assuming, given the state of the roads and the distances, that an average round trip would take two days, and that these trucks would work 300 days per year, this is a transport capacity of below 50,000 t/y, or about one sixth of the 1983 production. The private sector could have stepped in to offer road transport, but was itself facing shortages of trucks, spares etc. The distribution of cement to regions at a distance from the cement factories, more particularly in the North west and southeast of the country, has remained problematic throughout the 1980s, and with freight equalization abolished, both the availability and affordability of cement must have negatively affected demand in those regions. The subsequent economic liberalization has created an opportunity for the private sector to increase road transport capacity, and cement could now become more available, but not necessarily more affordable in those regions.

Cement prices were controlled up to July 1991 by the Price Commissioner, and the price level was always lower than the cost of production (Rwoga, 1993). The cement industry complained that this system contributed to their losses, because of the delays in applying price increases following rises of input costs, and because cost revisions often were lower than those applied for (Nyiti, undated). That notwithstanding, the cement price was increasing more rapidly than that of any other material listed by one source: by 10 times between 1982 and 1989, as to 9 times for steel, 5 times for corrugated iron sheets and only 4 times for lime, as shown in Figure 1 (UNCHS, 1991a; 1993).

It is unknown how the price liberalization, which happened after this period and allowed cement companies to set their own prices, has further affected those; but it is known that TPCC and TCC were making profits in 1992, and the losses at MCC were substantially reduced (Rwoga, 1993). The above increases refer to prices in Tanzanian Shillings. There seems to have been less variation as to the price of cement in US dollars, which consistently has been $100-110/tonne, but these are prices ex-factory and not what the end-user pays.

Lime

Lime is used in many regions of the country. Each category of user has its own requirements for quality: some want very pure lime, others want it very fine, again others very white, etc. Since the larger kilns generally make the better quality lime, it is their products which find a larger use in industry, e.g. in sugar refining or mining. And some industries, such as the paper and glass industry, simply make their own product according to their own specifications. Lime from the smallest kilns can be quite impure, as the more detailed information from Zanzibar points out (Bullard, 1988; Holmes, 1990; Holmes and Wingate, 1991; Haasnoot, 1993), and is therefore almost uniquely used in construction, where high purity is not a predominant requirement; in the case of Zanzibar, it has been suggested though to make certain improvements to lime production and processing, notably the introduction of lime putty, made by slaking lime under water; this is now being implemented by the Stone Town Conservation and Development Authority (Holmes, 1990; Holmes and Wingate, 1991). The intermittent kilns can produce lime of a reasonable quality; the one at Oldonyo Sambu for instance produced lime with a calcium hydroxide contents of 70%, which was considered as good as most lime available in the country (Spence and Sakula, 1980). Such kilns are

Table 7: Regional cement consumption in Tanzania, 1977-1994

REGIONS	1977	1978	1979	1987	1988	1989	1990	1991	1992	1993	1994
ARUSHA	18.26	15.48	13.71	23.04	14.76	20.56	14.44	15.28	25.7	36.05	43.8
COAST	12.24	19.7	2.62	9.45	7.41	7.41	4.83	18.93	3.42	1.35	2.51
DAR ES SALAAM	113.54	105.24	148.14	207.6	207.87	215.09	215.71	240.61	269.79	262.62	225.3
DODOMA	19.18	13.19	14.27	11.69	13.24	21.1	60.64	18.05	19.5	6.56	7.84
IRINGA	10.07	11.99	9.96	10.66	12.44	16.69	8.46	17.07	12.17	16.46	14.64
KAGERA	4.41	3.02	2.47	44.38	9.56	8.81	6.06	6.74	4.74	1.51	0.87
KIGOMA	3.37	3.45	2.33	4.69	4.46	3.26	3.52	4.65	5.16	2.56	4.14
KILIMANJARO	19.78	20.21	17.78	34.96	44.72	41.47	33.71	34.41	45.18	39.35	40.53
LINDI	7.25	4.31	4.3	2.26	2.33	4.29	4.63	4.1	1.98	0.63	0.54
MARA	6.12	6.45	5.53	12.83	6.07	9.07	4.27	10.57	6.54	3.77	2.41
MBEYA	13.51	14.24	28.73	28.06	32.23	28.03	31.25	38.15	26.59	36.79	41.78
MOROGORO	17.97	20.93	12.11	10.41	14.28	16.57	9.11	21.83	25.09	12.57	16.82
MTWARA	15.92	7.49	5.31	10.06	12.49	10.69	5.4	7.31	6.52	5.83	5.85
MWANZA	11.39	15.53	18.63	18.4	10.46	22.75	15.35	22.02	26.28	16.21	31.9
RUKWA	4.86	1.84	5.19	3.44	4.26	3.12	2.99	4.15	5.54	5.21	3.01
RUVUMA	6.96	3.03	5.64	4.81	4.26	5.15	3.18	5.98	4.23	4.27	4.04
SHINYANGA	4.13	3.92	5.79	8.75	7.95	5.16	8.42	9.21	10.08	5.1	9.39
SINGIDA	1.81	2.26	3.36	3.11	2.33	1.29	0.66	0.76	1.26	0.68	0.65
TABORA	4.96	6.55	6.71	7.36	7.78	7.49	4.31	4.22	4.55	4.18	6.78
TANGA	16.99	21.11	29.7	35.59	45.33	58.48	61.11	50.91	49.46	53.26	32.57
ZANZIBAR/PEMBA	2.51	1.67	6.33	17.85	25.66	31.88	30.7	32.36	35.76	36.79	42.19
TOTAL	315.23	301.61	348.61	509.4	489.89	538.36	528.75	567.31	589.54	551.75	537.56

Source: Tanzania Saruji Corporation   * Figures include Mtwara and Lindi

known to have supplied not only the building industry, but also road construction, the sugar industry and the mining industry.

Crushed limestone as road aggregate is probably used in several locations, but generally from quarries that only serve the road projects; there is only one case reported of a binder producer supplying road aggregate: this is the TCC, at about 5,000 t/y in the late 1980s (Austroplan, 1990). The large users of ground limestone are the Southern Paper Mills, in Mbeya region, and the glass factory at Dar es Salaam, who both have their own quarries and grinding plants, and National Chemicals in Dar es Salaam, who used to import it in the late 1980s, for fertilizer production. Aluminium Africa, in Dar es Salaam, was using 600 t/y in 1988, supplied by Tanzania Sandstone Ltd. There are other major users in agriculture and livestock, and minor ones for abrasives etc., some of which is imported. In all probability, it should not be necessary to import as much ground limestone as was reported in the late 1980s, unless the user industry has very specific requirements. But limestones of sufficient quality are available in the country, and grinding and milling plants, if not already available, e.g. in the cement industry, could be quite easily installed in a number of locations, to produce even the finest ground lime (lime flour); since most of the demand for this product is in Dar es Salaam, a location in that area, possibly to the north of the town, should be preferred.

Quicklime is mainly used in the paper industry, but produced by the paper mill itself, and for road building, although hydrated lime can also be used for the latter: it is less effective, but also less dangerous. A special kiln was constructed, at Kigamboni just South of Dar es Salaam, referred to as the Mji Mwema kiln in one source; it produced quicklime for a while in the late 1980s, but is no longer in operation. It is also known that, in the mid-1980s, a lot of lime had to be imported from Zambia, for the Songea-Makambako road in the South-West of the country. The distribution of lime to road projects can be quite problematic, because they create a sudden peak in the demand in a given region, which disappears again after two or three years. Roads do demand a lot of lime for stabilization, conservatively estimated at 20 t/km for Tanzania (Ikomba, 1994a), but elsewhere a consumption of up to 100 t/km is not unknown. It may therefore take the entire production of a substantial lime kiln, or the production of maybe 6-10 intermittent kilns, to build just one

Figure 1: Building materials price rises in Tanzania, 1982-89

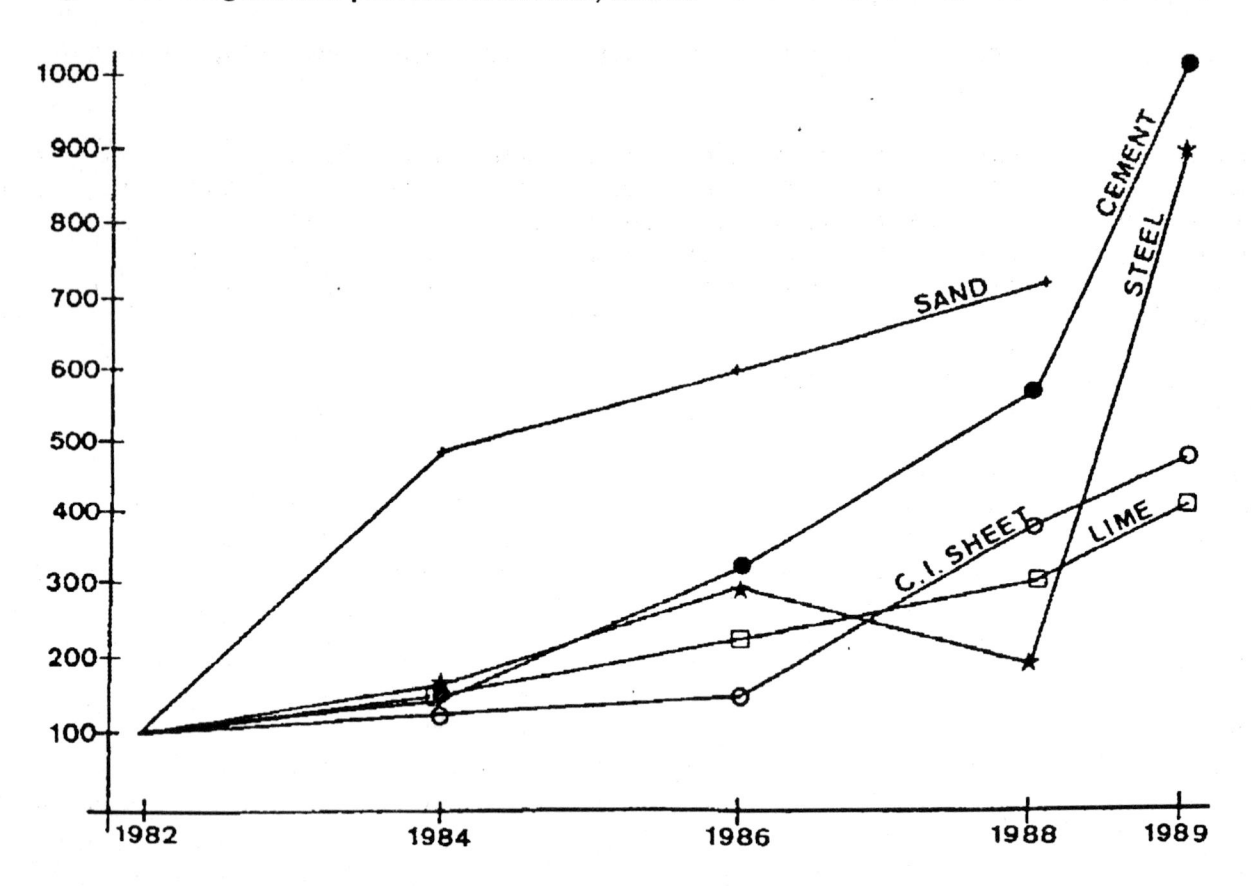

road. But it does not pay off to build these kilns just for the road project, if no other markets can be found afterwards, whereas existing limekilns that have the capacity to deliver the required quantities may be too far away to make that an economic proposal. Road construction and maintenance really need a network of lime kilns throughout most of the country, with some spare capacity to meet their demand.

The main use of hydrated lime is in building, particularly on the islands and along the coast, but it has been reported, to lesser extents, in other areas of the country. Building lime is currently mainly provided by heap burning in the informal sector of the economy. That makes it available very close to the main user groups, which saves on transport, and also able to react very quickly to increases or decreases in demand. It is furthermore provided by some intermittent kilns, and as a minor share of output by the large kilns.

Other important users of hydrated lime are the sugar industry, with refineries at Kilombero, requiring about 2,000 t/y, and Mtibwa, requiring about 600 t/y, both in Morogoro region, and others in Kilimanjaro and Kagera regions, each requiring about 600 t/y (Ikomba 1994a). Their main supplier used to be Tanga Lime, but since this closed about a year ago, Super Amboni Lime seems to have filled the gap. Tanneries in Moshi, Mwanza and Morogoro, each seem to require about 225 t/y (Ikomba,

1994a), and two major users in the mining industry, Williamson Diamonds (600 t/y) and Bockreef Gold Mining (120 t/y) were in 1988 supplied from Maweni Limestone in Tanga region (Austroplan, 1990). The paint industry is also reported to use around 600-700 t/y of hydrated lime, but some of that is imported (Austroplan, 1990). Given the location of the demand, it is quite astonishing that so much of this lime is provided from the Tanga area, when for instance lime kilns could be established in the Morogoro region to supply the sugar industry and tannery, or kilns closer by, e.g. at Dar es Salaam or Vitono, Iringa, to take that share of the market. And there also seems to be more potential for lime production in the north west of the country, with demands from a tannery, a sugar plant and mining, as well as in building in an area where cement is very expensive due to high transport costs.

Lime distribution and pricing has never been subject to as many controls and regulations as cement. The transport cost for lime would be the same as for cement, except that the industry probably has to rely almost entirely on road transport by the private sector, which is the most expensive form. The industry may have been faced with similar transport constraints in the past as the cement industry, but since lime is in average transported over shorter distances, these may have been less severe.

Compared to cement, lime prices in Tanzanian Shillings have gone up far less over the 1982-89 period, by only about 4 times (UNCHS, 1991a). Yet the sales price at the production site is still not much below that of cement, and the real advantage of using lime is therefore in areas where the transport cost of cement is high, whereas lime can be obtained locally, and otherwise in cases where cement is still in short supply. Prices of ground lime ranged from $7 to $44 in 1989 (Austroplan, 1990), but this is probably due to differences in quality, e.g. required fineness.

Quicklime was produced at $52 per tonne at Mji Mwema, whereas Tanga Lime sold hydrated lime at $61 per tonne in the same year. A later feasibility study (Ikomba, 1994a) put the sales price of the latter in 1993 at $67 per tonne. And in early 1994, Super Amboni Lime sold its lime at $65 per tonne, which was about two-thirds of the retail price of cement in Tanga at that time, being $95 per tonne; since then, the latter price has risen. Lime produced in village kilns is sold at around $48 per tonne, in 25 kg plastic bags (Ikomba, 1994b), which is substantially lower than the price of the big kilns.

Hydrated lime produced in heap burning varies in price; since packing is not in standard weight bags, any price given can only be estimative. Small bags sold on the beach just South of Dar es Salaam harbour went for approximately $70 per tonne in 1994 on site and $80 in retail, but there was no real competition of large kilns in the area in that period (Nguluma, 1994); lime from heap burning in Tanga was sold at TSh 500 per 25 kg bag, or $40 per tonne (Reuben, 1994). Comparative prices are reported from Zanzibar, albeit in an earlier period: of about $24 per tonne in bulk, and between $40 and 50 per tonne when bagged, in 1988 (Haasnoot, 1993), which partly explains the popularity of lime on the island. Transport costs can put up the price of lime to the user considerably; hydrated lime delivered to Williamson Diamonds in 1989 arrived at $185 per tonne, or three times its cost price off the producer in Tanga region (Austroplan, 1990).

With a few exceptions, lime production in Tanzania is currently concentrated in the coastal belt and on the islands. In the case of the islands, there is a long tradition of building with lime, the material is available relatively cheap and without huge transport costs, and cement is more expensive and harder to get; it is logical that lime flourishes in those circumstances. But on the coast, conditions are not the same. The southern coast is far from the cement factories and has some tradition of lime construction, and therefore there is a flurry of heap burning with a ready market. In Dar es Salaam, there is competition from the cement plant, and there are no major lime producers currently operating, which has given opportunities to some heap burners. There are, however, several industries in Dar es Salaam and even as far as Morogoro, which require lime or limestone products, which the region could provide.

There is major, but currently somewhat under utilized capacity of lime production in Tanga region, which traditionally seems to supply many other regions of the country. Given the transport costs involved, this does not always makes sense, and one would expect more lime production to happen in the centre, north and west of the country; given the current climate of economic liberalization in the country, this might just happen in the coming years. However, there is less of a tradition of building with lime in those areas than along the coast and on the islands, and the take off of a building lime market may therefore be slow.

Pozzolanic binders

Lime-pozzolana was only ever produced at a very small scale at Oldonyo Sambu, in Arusha region. It achieved 28-day strengths of 2.0 N/mm^2 on 1:2:6 lime:pozzolana:sand mortar cubes, and 3.0 N/mm^2 in prefabricated blocks, which were considered adequate qualities for simple building purposes, such as housing (Spence and Sakula, 1980; UNCHS, 1993); a demonstration building still stands on the site, but has suffered some erosion on its most exposed surfaces. It was reported at the time, that the retail price of lime-pozzolana was about 60% of that of cement (Sakula, 1980c), which should have made it an interesting alternative. But, this binder only managed to reduce the cost of local housing by about 25% compared to Portland cement; and those houses were still twice as expensive as the ones using the traditional mud and pole, and few of the inhabitants were able to afford the material. In reality, therefore, it was more aimed at a different market, which might have required a different management structure (UNCHS, 1991).

Tests on PPCs conducted at the University of Dar es Salaam showed that blends containing 20% pozzolana met the requirements of British Standard 12, and so were suitable for ordinary concrete works, whereas blends containing 40% pozzolanas would be suitable for low grade cements. As pozzolanas do not have to be clinkerized, but can be interground with the cement clinker and gypsum, which would normally save on energy (unless the particular pozzolana is a very hard material), this could reduce cement costs, by as much as 20% for a PPC containing 40% pozzolanas, and 7% for one containing 20% pozzolanas (Stewart and Muhegi, 1989). Pozzolanas are not available near every cement plant, but the one at Mbeya has its own quarries, which it recently started to use for the production of a PPC.

Gypsum

This is produced in small quantities in Moshi, but largely used by the producer itself, for manufacturing school chalks. It is unknown what the material would cost in retail.

Figure 2: Cement production per capita in some Third World countries (UNCHS, 1993)

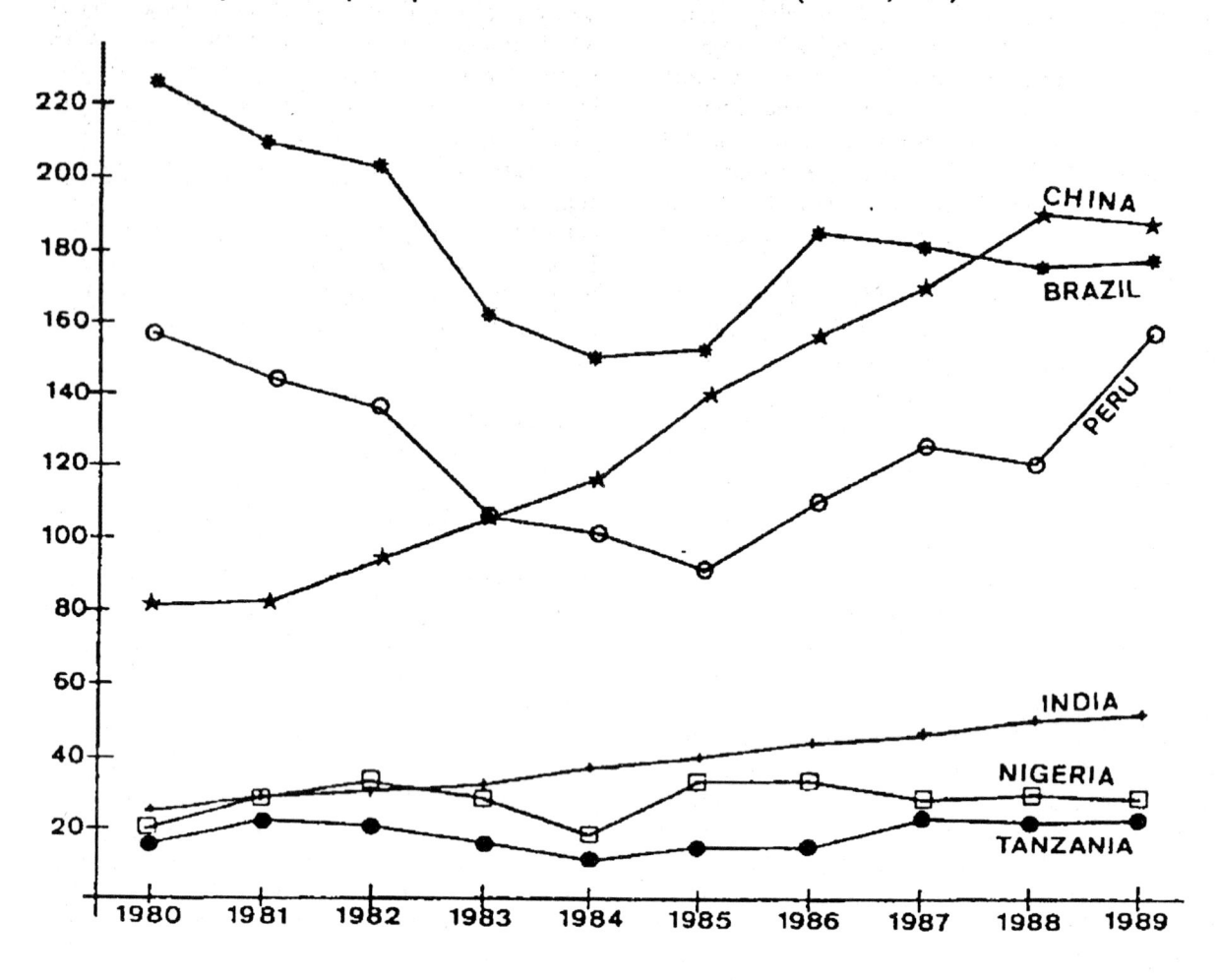

Future trends

Cement

As mentioned previously, we do not have sufficient details about the current use of cement for various building purposes and other works. We do know that the annual expenditure on housing is much higher than that on other construction, and that its informal sector share is much larger than its formal sector share (Karlsson et al, 1981). That does not necessarily say much about cement consumption, since the same source admits that the vast majority of houses, particularly in rural areas, only uses local materials, and has an average lifespan of seven years.

The figures from the census of 1977, compared to those of 1967, do, however, bring out tendencies (Salomon, 1980). There is a noticeable increase of the use of foundations in housing, from 10 to 27%, and the proportion of foundations with cement, concrete or bricks has risen from 1 to 6%. And the use of cement floors has increased from 11 to 15%. So there is some modernization of housing, probably largely linked to urbanization. In fact in 1977, 60% of urban households used cement floors, versus 6% in rural areas. The increased urbanization of the country may well be a major factor behind the increased per capita use of cement in subsequent years, from 17.2 kg per capita in 1978 to 22.3 kg per capita in 1988 and possibly 22.8 kg per capita in 1992; where the urbanization rate was 12.7% in 1978, it had reached 18.3% by 1988 and was predicted to reach 25-30% by the year 2000 (Kulaba, 1980; Tanzania Bureau of Statistics, 1988).

There is a noticeable worldwide tendency for cement consumption to be linked to development in general, and income levels in particular. The current cement consumption in Tanzania is clearly very low still, in comparison with other Third World countries (Figure 2), where medium-income countries like Brazil or China had reached levels of 150-180 kg per capita in 1989, and India had reached about 50 kg per capita, about double its level in 1980; over that same period, Tanzania's consumption increased by only about 30%. The average increase in per capita cement consumption in Tanzania, over the 1978-1992 period is 0.4 kg per capita per year. If this trend is to continue unchanged, by the year 2000 consumption would stand at 26 kg per capita, and the national cement consumption at around 840,000 tonnes per year.

But will the trend continue unchanged? Stewart and Muhegi (1989) estimated the real demand in

1987 at 40 kg per capita, or 920,000 tonnes for the country as a whole, in fact implicating a continuing shortage of around 40%. At the same time, the TSC was using targets of 50 kg per capita in its second corporate plan (1981-86), with a demand increasing at a rate of 7% annually, which may have been more a case of wishful thinking than of an actual informed market study. It is clear that actual consumption does not equal demand, largely because it is dictated by the quantity of cement on offer. With relatively low levels of capacity utilization in the cement industry, substantial exports, and no imports to speak of, there may be some unsatisfied local demand, but whether this stands as high as the above estimates is doubtful. Factors in favour of an increase in demand, beyond the current level of growth, include the economic liberalization, which will hopefully lead to more initiatives, and with that more investment in building and to a reduction of some of the traditional bottlenecks in supplies of imported fuel, equipment, etc., as well as ongoing work on improving the country's infrastructure, particularly roads, which will diminish transport problems. Factors against may be that the bulk of the population may end up with less to spend, and a slow increase in competition of cheaper binders, particularly lime.

It is very difficult to make correct predictions of the development of cement demand over the coming years. From available data, we assume that there are four major factors influencing this demand:

(1) Transport costs, influenced by available means of transport and distance from the consumer to the cement factories. Table 7 shows clear differences in regional consumption of cement, with outlying regions having very low consumption levels.

(2) Economic development, as expressed by increases of the Domestic Product and investments, or of incomes. During the late 1970's and most of the 1980's, Tanzania's economy did not do well, but following the Economic Recovery Programme introduced in 1986, it has started to pick up, and the economic growth stood at 3.3% in 1989 and 3.6% in 1990, which is exceeding the population growth of then 2.8% (Nyiti, undated).

(3) Population growth, which has been steadily rising since the second world war from 3.0% per annum in 1948-57, to 3.1% in 1957-67 and 3.3% in 1967-78 (Kulaba, 1980) when it peaked, now has started to decrease, currently reaching around 2.8%. In absolute terms, this translates into an increase of about 750,000 people this year, and at around 25 kg of cement per capita, that means an additional demand of 17,500 tonnes per annum.

(4) Urbanization, which influences cement demand because towns are centres of investment, and also because building standards and regulations apply, which require a certain quality of construction, causing a much higher level of cement consumption than in rural areas. Urban population growth in Tanzania has been consistently much higher than the national average, and stood at 7.4% per annum in 1948-57, 6.5% in 1957-67, and peaked at 9.2% in 1967-78; it has slowed down since then, to about 7% over 1978-1988 (Kulaba, 1980; Tanzania Bureau of Statistics, 1988).

In 1992, local consumption of cement reached a peak, at 589,540 tonnes, or 22.8 kg per capita, to subsequently drop again in the next two years. In fact, since about 1987, the per capita consumption of cement has not markedly increased, notwithstanding the ongoing urbanisation and the better economic performance of the country later in that period. A bottom-line planning of cement demand by the year 2000 would therefore only take a population increase, of 2.8% per year, into account, and arrive at approximately 736,000 tonnes of local demand, growing with about 21,000 tonnes each subsequent year. A more optimistic view would be that the economic liberalization and increased urbanisation would start to have more impact, and that per capita consumption would start to rise again, possibly with the previous rhythm of 0.4 kg per capita per year (or about 1.7%); this would produce a total annual growth in local demand of 4.5%, well below the TSC prediction, and lead to an overall local demand of around 840,000 tonnes by the year 2000, growing with about 38,000 tonnes each subsequent year. From all the previous trends and developments, it would appear unrealistic to plan for demands of 40 or even 50 kg per capita per year, as suggested in the early 1980's.

The current nominal capacity of the three factories stands at 1,270,000 tonnes per year. However, capacity utilization has not been good in the past, and the plants are not in the best condition; it will take time and effort to get closer to the high level of capacity utilization (80-90%) that is for instance common in Kenya. A report dated 1992 estimates the realistic capacity for that year as being in the 900,000-1,000,000 tonne range (UNIDO, 1992). That would still be adequate to cover local consumption up to the year 2002 at least, even in the higher consumption scenario, although exports would slowly start to suffer.

The TSC is aiming at a partial privatization of the industry, bringing in fresh capital and expertise to enable a major overhaul, and a substantial increase in capacity utilization. This was achieved at the Dar es Salaam plant in the mid 1980's, by Scancem, who became a minority shareholder; the capacity utilization at that plant had reached around 80% by 1993, and it is unrealistic to expect an improvement to above 90%. Danida is assisting the overhaul of the plant at Tanga, over the 1992-1994 period, which should raise its capacity utilization from about 50% in 1993 to possibly around 80% as well. The plant in Mbeya is a major concern: it has never produced at more than 40% of its rated capacity, and never made profits. And its potential nearby market, in Mbeya and surrounding regions, is relatively small. The plant is trying to attract partners, and its survival may depend on the recent change towards producing Portland Pozzolana cement (Nyiti, undated).

As early as 1989, Stewart and Muhegi argued for the establishment of additional capacity for cement production in the form of vertical shaft kilns, of a capacity of around 60,000 t/y. They explained that these would be more economical than rotary kilns up to a capacity of 750,000 t/y (excluding distribution costs) or 1,000,000 t/y (including such costs to adjacent regions). Such shaft kilns could bring cement production to many outlying regions, and would only require limestone deposits of at least 3.6 million tonnes, of which there are many in the country. They would substantially reduce the distribution costs of cement to such regions. Both foreign and local private companies have shown interest to set up such mini cement plants, e.g. in the Lake Zone or in Shinyanga (Nyiti, undated), but these plans have not yet materialized. If the current capacity of cement production has to be increased in the near future, as seems certain from the above projections, it would make sense to consider mini cement plants for an extension of the capacity, in the first place in the North West of the country, and possibly also in the South East.

Lime

To estimate the development of the demand for lime is as difficult as for cement. A large part of the current demand for lime is for construction, particularly housing. But it is known that, the more a country develops, the more lime it needs for other purposes, particularly industry. Tanzania's economy is growing at the moment, with 3.6% in 1990, and it is estimated that growth will attain 4-5% per year in the 1990's (Ikomba, 1994a). It is quite likely that the demand for lime will grow with similar figures, which is in line with the more optimistic growth scenario for cement demand.

The demand for ground limestone, for various purposes including glass and paper manufacture, chemicals, agriculture, livestock and metallurgy may rise from its current level of at least 20,000 t/y to about 25,000 t/y by the year 2,000. About a third of this will be produced by two large consumers themselves: the glass and paper industries, and they have ample capacity to meet their own demand. That leaves about 17,000 t/y to be delivered by the local lime industry, or to be imported. The local industry should be able to deliver the bulk of this amount. This excludes a demand for agricultural lime, which reportedly is not used at the moment. Certain lime producers might be able to make a more efficient use of their waste products, by offering a range of ground and milled limestones, including agricultural lime, for which an occasional market might be found if it is sufficiently cheap. The other potential market for waste materials to be explored much more actively by lime producers is that of aggregates, both for road works as well as concrete used in building, and it is thought that the market for crushed limestone can expand considerably.

Quicklime is currently mainly used for papermaking, where demand might increase to about 6,000 tonnes by the year 2000 (equivalent to 8,000 tonnes of hydrated lime), and in road construction. The latter market is very variable, depending on where and when road projects are taking place. Its demand was estimated at 2,700 tonnes in 1992 by one expert, and could increase to 4,000 tonnes by 2,000, or 5,300 tonnes of the equivalent hydrated lime. The lime for papermaking is produced by the Southern Paper Mills itself; the product required for roads needs some grinding after firing, and can therefore only be delivered by the larger producers, who have the required equipment.

The use of hydrated lime outside building is estimated to increase from 4,000 to about 6,000 tonnes by 2,000 in the sugar industry, and with the same amounts in the remaining industry, including tanneries, mining, paint etc. This lime needs to be of good quality, and is therefore mainly supplied by the larger lime works. The use of hydrated lime in building, currently estimated at 40,000 tonnes, may well increase to about 55,000 tonnes by the year 2,000. These figures will have to be confirmed by further research. It is thought that the use of lime, particularly in housing, will increase because of its cost advantage compared to cement mainly thanks to its production close to the user. This lime does not necessarily have to be of as high a quality as the industrial lime, and can therefore easily be produced by much smaller kilns, such as the heap kilns, which may well continue to play a very important role. A large lime kiln could produce such lime, possibly at nearly comparable prices, but would have a much larger distribution radius, which would put the lime price up away from the factory, and reduce demand.

In summary, there will be an annual estimated demand for hydrated lime equivalent of about 80,300 tonnes by the year 2,000, about 25,300 tonnes of which needs to be of relatively high quality.

The current capacity of the lime industry is much beyond that, as explained previously. Large kilns have a current annual capacity of around 66,600 tonnes, and there are another 3 kilns on the drawing board or under construction, bringing that total to 99,600 tonnes. The intermittent kilns have a probable capacity of around 4,000 tonnes per year at the moment, but if the abandoned kilns were to be overhauled, this amount could well triple. And there is potentially no end to heap burning capacity. The situation of the large kilns in particular is of some concern: there seems to be an overcapacity, certainly if the three new kilns come into production, because not all of them will have much of a market for building lime. Having two large kilns near Tanga, far away from most industrial and road markets, and with a limited market for building, seems too much for instance. On the other hand, there is no sizeable kiln in Morogoro region, where demand is quite high. And there still is a substantial number of potential lime users in the country, outside the islands/coast/northeast, not located within easy reach of a lime kiln, who would benefit of a network of small to medium-size kilns very close to these markets.

Pozzolanic binders

At the moment, there is no production of lime-pozzolanas, but it would not take much to establish it. The plant in Oldonyo Sambu still has a working lime kiln, and with the very fine pozzolanas of the area, which do not need milling, lime-pozzolana production could be re-established there relatively easily. As mentioned earlier, lime-pozzolana was produced at Oldonyo Sambu on an experimental scale, in the late 1970's. After handing over, the plant soon ceased production. There have been several attempts since to revive it, including an ECA proposal of 1983 to make it a demonstration project for the East African region, but none of these materialized. The evaluations carried out after the project ceased, suggest that the material tended to largely benefit users outside the rural community, and that a more commercial approach and a different management structure might have been required (UNCHS, 1991). The material has been used on a couple of buildings, and some of the prefabricated blocks made with the binder and pumice as aggregate show considerable erosion, after 15 years of use on very exposed parts of the buildings, such as corners or wall bases. Further research will be needed into the most appropriate uses of this material, as well as mortar compositions etc. The same project had started research on the production of lime-pozzolanas using other pozzolanas within the country, with encouraging results for some of the South Western volcanic pozzolanas, as well as rice husk ash; there is also the potential of using fired clay or bagasse ash. Experiences in other countries, such as India, and to some extent in Oldonyo Sambu, show that it is technically feasible to make satisfactory binders for certain purposes, such as masonry or renders, from lime and pozzolanas. Tanzania has ample potential to produce such materials, but this remains sofar untapped.

The University of Dar es Salaam has done research on Portland Pozzolana cements, substituting 20 and 40% of the cement clinker by Tanzanian pozzolanas; in the first case, a cement was produced which could compete with ordinary Portland cement in quality; in the latter case, a binder of lesser strength was produced, which was still perfectly adequate for many construction purposes (Stewart and Muhegi, 1989). The cement industry in Tanzania has now just started to produce a PPC in Mbeya, something that is quite common already in other countries. The industry is fully aware of the advantages of producing PPC instead of OPC, for instance: savings on fuel and on finite limestone resources, a reduction of environmental damage, and certain qualities of the binder itself. A UNIDO project supporting the development of the cement industry in the PTA region has expressed this repeatedly (UNIDO, 1992; 1993), and even went as far as suggesting that the survival of MCC may well depend on a shift towards PPC manufacture. It is less likely that it would be economical for TPCC or TCC to produce PPC, given their considerable distance from adequate sources of pozzolanas.

The exact composition of the PPC currently produced at Mbeya is unknown to the authors. But it is generally accepted by various standards organizations that Portland Pozzolana cements should not contain more than 30%, or exceptionally 40% pozzolanas, to keep these cements at a comparative quality level to Ordinary Portland cements. Few construction purposes, however, require these high levels of quality. The real need in the Third World is not for a high-strength, fast-setting material. Most buildings, such as houses, clinics and schools, are relatively small, are not built in a hurry, and do not require excessive strength. The United Nations Centre for Human Settlements, UNCHS, estimates that only 20% of the worldwide use of cement requires the strength of an OPC (UNCHS, 1989). This is repeated by Stewart and Muhegi (1989), who go on to mention that another 40% has intermediate strength requirements, and the remaining 40% used for applications such as mortars, plasters, foundation concretes, concrete blocks and soil stabilization, where low grade cements could be used. At the moment, the TPCC occasionally produces small quantities of masonry cement (UNIDO, 1993), undoubtedly not using pozzolanas, but there is clearly a much greater scope for a range of cements to be produced, some of which would be OPC or PPC for high standard applications, but others could include much greater percentages of pozzolanas, in combination with cement clinker or lime.

Development Issues

Access to binders

The consumption of binders in Tanzania is not high. People currently use an estimated 2.2 kg of lime per capita per year, and around 22.8 kg of cement. And in several outlying regions, cement consumption is a lot lower, from 2.3 to 7.0 kg per capita. Access to binders is mainly influenced by two factors: affordability and availability.

Up to 1982, there was a freight equalisation charge on cement, which roughly unified its cost throughout the Republic, and favoured the more remote areas. Between 1982 and 1989, the cost of cement in TSh increased tenfold, which is probably more related to the bad performance of the industry than to the abandonment of the freight equalisation charge; this increase was bigger than for most building materials, and by far exceeded the increase in the cost of living index. The cement price is currently around $100-110 per tonne ex-factory. It is available in retail, close to the factories from about $120 per tonne, or $6 per 50kg bag. But in far away regions, such as Kigoma, it may be as expensive as $400 per tonne, or $20 per bag in retail, which equals 40% of the monthly minimum wage. Cement is therefore much more affordable close to Dar es Salaam, Mbeya and Tanga, than in more remote areas. It is unknown whether the PPC now produced at Mbeya is actually cheaper to the user than OPC.

Lime prices at the production site vary much more, from around $36-50 per tonne amongst the smaller producers, to around $65 per tonne at the bigger plants, who sell a superior product. When lime is produced by heapburning or in small kilns, the production site is often also the retail site for the surrounding area, whereas the large producers have to rely more on retail via shops, which can make their product almost as expensive as cement. But where there is a decentralized network of small producers, lime can be substantially cheaper than cement, often much less than half its price, which partly explains its popularity in Zanzibar. This is important for the building market, because if lime is used instead of cement in mortars, plasters etc., the dosage in binder of those mortars should be increased, probably by 50-100%, to achieve similar results; and lime therefore needs to be a lot cheaper than cement, to be able to compete. This seems feasible with very small scale production methods even relatively close to the cement industries.

There is no current production of lime-pozzolanas, but when they were produced near Arusha in 1980, their cost arrived at about 60% of the cost of cement, and it was claimed they could reduce housing costs by about 25%. This certainly is a positive development, but whether it is really enough for the low-income groups, remains doubtful.

The availability of binders is influenced by the distance to its production sites and above all by the transport infrastructure. Cement distribution throughout the years has been greatly hampered by inadequate rail infrastructure, which is the cheapest way of transport, but until recently also by a shortage of trucks. It has therefore always been difficult to supply regions further away from the factories, which has raised a certain interest in the establishment of mini-cement plants, for instance in the North West. The small-scale lime industry does not suffer from similar problems, because it only serves restricted markets in the vicinity of the plants. But the medium- to larger scale lime works suffer similar transport problems, which may be the most acute when large quantities of lime are suddenly needed for road works, but could also apply to some of the larger industrial users.

Transport

Binders, like many other building materials, are quite bulky and heavy, and therefore claim a lot of transport. The cheapest ways of transporting them are by rail or boat. The latter method is used by the TCC, to supply Lindi, Mtwara and the islands. The capacity of the railways, though, to transport the required quantities of cement is very insufficient, as low as 20% of requirements in 1981. Road transport is twice as expensive as rail transport on tarmac roads, but may cost up to 7 times as much when dirt roads have to be used. It has also traditionally been insufficient, which was one reason to establish the Saruji Trucking Company; but that company also managed to transport only one sixth of the required tonnage in 1983. The result is that outlying areas receive less cement than what they might want. With the expected improvement of capacity utilization due to past and ongoing maintenance, transport may become more of a bottleneck, unless the private sector is encouraged to step in increasingly.

Transport is not only a bottleneck in distributing the finished product, the lack of it can also hamper raw materials supplies. TCC and MCC have both experienced problems with their quarrying and haulage equipment, and raw materials transports have also occasionally posed problems to the larger lime or lime-pozzolana industry. As a general rule, the smaller producers suffer the least transport problems, because they can often use alternatives, such as oxcarts, donkey carts, small boats, small vans or trucks, or even wheelbarrows, which stimulates the local economy, and is less of a drain on the national one.

Energy and the environment

The production of binders requires considerable quantities of energy, and it is important to do this as fuel-efficiently as possible; but the choice of the type of fuel is equally important, in terms of the national economy, the environment etc. Besides, it is probably correct to consider energy use in terms of its overall use in building, which would include energy used in production and in transport, and take account of replacement ratios.

The TCC uses fuel oil for its kiln and electricity for its grinding mills, at a rate of 4-4.6 MJ/kg cement, which is satisfactory for that type of cement factory. The comparative figure for the MCC, which also uses local coal, is 4.6-5.4 MJ/kg. Tanga Lime is reported to have used fuel oil at about 5.2 MJ/kg lime, which is not bad compared to the much larger cement kilns, and could be further improved by installing oil burners which better distribute the fuel throughout the kiln. But the wood burning lime industry uses a lot more energy: around 15 MJ/kg for the shaft kiln near Arusha, and as much as 35 MJ/kg in heap burning. The latter production method is particularly inefficient in energy terms, which again has a negative impact on the environment.

A disadvantage of using fuel oil is that it is expensive. Another major disadvantage is that it has to be imported, and therefore requires foreign exchange; this has caused problems in the past, and even nowadays, with the industry allowed to export and retain some of the foreign exchange gained, this is insufficient to cover fuel imports, and should preferably be reserved for other inputs, such as spares, equipment and management skills. The Mbeya factory is trying to use the coal of that region, which enables a saving of about 30% on the fuel bills, but its supply has been irregular and the coal is not very pure.

The smaller-scale lime industry largely uses timber, and some agricultural waste; this may be the equivalent of 100,000 trees annually, which has raised concerns, for instance for the survival of mangrove forests in some areas. A shift from heap burning to small kilns could save at least half of that energy, but may not always be feasible. Wood is a renewable resource, and there may be some merit in looking at reforestation for lime burning purposes; in addition, there may be more scope for the use of various waste materials as fuel.

The cement industry, and some of the larger lime plants, also use electricity, particularly for grinding, but also for other purposes. In some years, such as 1992, drought has badly affected electricity supplies, reducing the production capacity of those industries.

The overall use of energy in building with binders would also have to incorporate the transport energy. Figures for India put energy use for transport by truck at 2.85 MJ/tonne/km, 0.9 MJ/tonne/km by rail and 0.09 MJ/tonne/km by sea (UNCHS, 1991b). In the case of Tanzania, these figures are probably somewhat higher, particularly for road transport. In the latter case, the energy requirement of cement brought to site may well be twice its production energy, once it is transported over about 500 km. Taking transport into account, it would be more favourable in energy terms, to have a decentralized network of smaller production units, including mini-cement plants and alternative binders.

For lime, it will remain hard to compete with cement in energy terms, even when large kilns are used and there is some advantage in having less transport; this is largely due to the replacement ratio in its use.

The incorporation of pozzolanas in binders, though, can lead to dramatic savings. Pozzolanas require much less energy, and sometimes hardly any at all, to be turned into a binder; the volcanic ash from Mount Meru, for instance, did not need any processing. A 1:2 lime-pozzolana, produced at the small plant of Oldonyo Sambu, would cost about 5 MJ/kg in energy, which is in the range of binders produced by large plants. The incorporation of pozzolanas in PPCs can also reduce the energy use in cement production, often by as much as 20%. In terms of energy savings, pozzolanas deserve a lot more attention in Tanzania.

The type and quantity of energy used have a clear impact on the environment. But there are other factors involved in that as well. The most important one is the quarrying of limestone. Uncontrolled quarrying of some reefs and small islands in front of the coast for heapburning of lime, have raised concerns over coastal erosion. Quarries also often leave gaps in the landscape, without any subsequent use, particularly in the case of informal sector production, and more attention is needed to make disused quarries useful, for instance as fishponds or for agricultural purposes with some refill. Finally, quarry waste can be a problem, which requires further exploration of secondary uses of crushed or milled limestone, and quarry management that takes not only binder production but also other markets into account.

As a positive contribution to the environment, certain waste materials could be turned into binders, but this is not actively pursued at this moment. Agricultural waste, such as rice husks, can be made into pozzolanas, and the same applies to some of the (underfired) waste from clay brick or tile production. Other agricultural wastes may be turned into fuels, and this happens to a limited extent with, for instance, coconut shells; but the potential could be explored more systematically.

Manpower

The productivity in the cement industry is not high, compared to other countries; labour force motivation has been quoted as a factor in this. The TCC required 8.8 manhours to produce a tonne of cement in 1982, but this was lowered to 4 mh/t by 1993, due to an improved capacity utilization. This compares to about 3 mh/t for similar factories in India, and 0.3-1.0 mh/t for similarly sized but more mechanized factories in the North. At the MCC, those figures have at best been 12.7 mh/t.

Lime production uses a lot more manpower: Tanga Lime was reported to have used 50-60 mh/t around 1990, but only around 31.5 mh/t back in 1982, and Mji Mwema 34.6 mh/t; the village kiln at Ikengeza was thought to require 16-20 mh/t, but that figure seems too low. Heapburning is estimated to use 60-90 mh/t, unless explosives are used in quarrying, when it may drop to around 50 mh/t. It is therefore safe to assume that most limeburning in Tanzania requires in the range of 40-80 mh/t. In other words, the production of lime creates 5-10 times more jobs per tonne than cement, depending on what end of the scale we look at.

It is also likely that productivity can improve across the board, by better capacity utilization, higher motivation of staff, training, and improved management. Within the cement industry, the Saruji Training Centre caters for in-house training of artisans and technicians, but it would need upgrading to deal with higher level staff. The cement industry has particularly suffered from a lack of local management expertise, which in some cases has been tackled by bringing in a foreign company. In the case of lime production, there is no provision for formal training; all skills are transferred on the job, and there is obviously some scope for improvement.

Capital investment and foreign exchange

Investment data for the various technologies of binder production is incomplete. To start heap burning of lime requires no more than $1,500 for a sizeable operation, or around $100 per created workplace, and it will often be much less. The smaller lime kilns cost in the order of $2,400-$4,000, which is at least two to three times as much per workplace, and the large oil-fired kilns come in the $20,000-60,000 range, which is at least five times as much per job created, or somewhere in the range of $400-600. The current value of the TCC is estimated at 5.2 million $, or about $9,300 per workplace, and of the MCC at $7.2 million, or about $12,200 per workplace, but these are figures after depreciation, and the real figure at the time of investment must have been substantially higher, possibly as high as $25,000, or 40-60 times the investment per workplace in a large lime factory.

It is quite clear, therefore, that investing in lime has several advantages over investing in cement. The employment generation potential is a at least forty fold. Also, it is much more easy to get access to smaller amounts of capital, which allows small- to medium-scale entrepreneurs to enter into production; on the other hand, it may still be difficult for the informal sector, mainly the heap burners, to move up in scale, because of credit barriers to that sector. Also, an investment in lime production produces a much faster return on capital than one in cement, where it has taken ten years to get two of the factories going.

Another problem is that the cement factories, and to a limited extent the larger lime plants, do require substantial amounts of foreign exchange for their installation, for spares and equipment, fuel, and sometimes foreign skills. In some cases, this attains more than half of the running budget. The cement industry is now allowed to retain foreign exchange gained through exports, and this helps to some extent with the imports, which in the past have been a major bottleneck. Still, they do not export enough, and exports are slowly decreasing. Foreign exchange is therefore likely to remain a serious constraint for the cement industry, and possibly for the larger lime plants.

Conclusions and recommendations

- Transport is a major bottleneck in a country like Tanzania, with a relatively low population and infrastructure density. In order to reduce the transport component of the binder cost, it would be advisable to plan future expansions of the industry in a more decentralized manner, incorporating possibly mini-cement plants and a network of lime or lime-pozzolana production units.

- At the same time, the establishment of smaller production units is also favourable in terms of employment generation and savings on capital and foreign exchange.

- Tanzania should attempt to make better use of its pozzolanas, in order to reduce the energy requirements in binder production, and to reduce the costs of binders. The MCC could largely produce Portland Pozzolana Cement, and there is ample potential for producing lime-pozzolanas, although more research and development is likely to be needed.

- The increased use of pozzolanas could also contribute to a diminution of certain waste materials such as rice husks or bagasse ash; some waste materials can also be used as fuels. The use of waste materials in binder production is relatively limited and deserved to be explored and promoted more systematically.

- It would be advantageous for the country to have binders with different qualities for different purposes, and to develop building standards and regulations accordingly. These could for instance include:
  - high-strength cements: OPC and PPCs with up to 30% pozzolanas,
  - medium-strength cements: masonry or blended cements,
  - low-strength cements: lime, lime-pozzolanas and some blended cements.

- The development and dissemination of new types of binders, or the revival of old ones such as lime-pozzolanas in Zanzibar, requires a substantial amount of communication and training, for which adequate capacity is lacking in the industry; this will require external support.

- The heapburning of lime has serious environmental consequences, particularly due to the large amounts of fuel used. Heap burners need support for a transfer to more fuel-efficient lime production, possibly requiring the introduction of small kilns, training and access to small credits.

• Every attempt should be made to reduce the imports of fuel-oil for cement or lime production, by substituting it with local coal or plantation firewood, and occasionally waste materials.

• The binder industry needs to develop secondary markets for by-products made from waste limestone, including aggregates for road construction and concrete, ground lime for agricultural purposes, cattlefeed or industry, etc.

• Disused quarries need to find secondary uses, as fishponds, reforestation sites, etc.., and quarry owners have to be stimulated or, if needed, forced to implement quarry re-use.

• There has to be better coordination between road planners and the lime industry, to enable the latter to plan for sudden increases in production caused by peak demands from road building.

• There is some concern about the health of workers in the binder industry, and more particularly caused by dust in lime production; these health issues deserve more attention.

References

W.J. Allen: "The evaluation and testing of a volcanic pozzolana", PhD Thesis, 164 pp. + appendices, 1981.

Austroplan: "Development of Lime Production in the SADCC Region", United Nations, New York, 1990.

P.A. Bengtsson: "Tanzania Saruji Corporation in Relation to National Housing Policy", paper presented to the Conference: *Towards a National Housing Policy,* Arusha, 10 pp., 1980.

Blue Circle Group: "Selection and Design of cement plant equipment: The technical aspects of scale", 1982.

A.W. Bullard: "A report on lime production methods on Zanzibar Island and the use of Appropriate Technologies to improve efficiency and quality", ITDG, Rugby, 1988.

P. Cappelen: "Pozzolanas and Pozzolime", Working Report 17, Building Research Unit, Dar es Salaam, 26 pp., 1978.

Daily News: "Tanga Cement Plant: Success Story", Dar es Salaam, 25 March 1982.

J. Haasnoot: "Small-scale lime burning in Zanzibar: Survey of present activities -Proposals and recommendations for improved technology", STCDA, 20 pp. + appendices, 1993.

N. Hill: "Energy and Cementitious Materials: a burning problem", pp. 11-17 in *BASIN News* No.6, St. Gallen, July 1993.

S.D. Holmes: "Report on the Proceedings of the Zanzibar Stone Town Workshop and Advice to the STCDA on lime technology and historic building conservation", ITDG, 24 pp. + appendices, 1990.

S.D. Holmes and M. Wingate: "Report and advice given to the STCDA for emergency repairs, traditional lime technology and small scale lime production", ITDG, Rugby, 65 pp. + appendices, 1991.

E.G.S. Ikomba: "A case study of small scale lime production: Kigamboni Limeworks, Dar es Salaam", Dar es Salaam, 5 pp. + appendices, undated.

E.G.S. Ikomba: "Feasibility Study for lime and precipitated calcium carbonate", Lime Consult, Dar es Salaam, 45 pp. + appendices, 1994a.

E.G.S. Ikomba: "Lime and lime pozzolana units in Tanzania, a case study", Dar es Salaam, 7 pp., 1994b.

E.G.S. Ikomba: "Village lime kilns: SIDO Experience, a case study", Dar es Salaam, 5 pp., 1994c.

R. Karlsson, R. Salomon and I. Siroiney: "Requirements of rural housing and building materials: some basic facts and figures", pp.26-33 in T. Schilderman (ed.): *Rural Housing in Tanzania: Report of a seminar organized at Arusha,* 1991.

R.H. Kimambo: "Development of the non-metallic minerals and silicate industry in Tanzania, vol.II", 1988.

L. Komba: "The Population of Tanzania", paper presented to the Conference: *Towards a National Housing Policy,* Arusha, 3 pp., 1980.

S.M. Kulaba: "Housing, socialism and rural development in Tanzania: a policy framework", paper presented to the Conference: *Towards a National Housing Policy,* Arusha, 62 pp., 1980.

J.F. Martirena Hernandez: "The development of pozzolanic cement in Cuba" in *Appropriate Technology* Vol.21 No.2, Sept. 1994.

Ministry of Lands, Housing and Urban Development, Surveys and Mapping Division: "Atlas of Tanzania", second edition, 1976.

Ministry of Tourism, Natural Resources and Environment: "Management Plan for the Mangrove Ecosystem of Mainland Tanzania", Dar es Salaam, 1991.

B. Muhegi: "The need to promote the development of Minicement plants and cementitious materials in Tanzania", IHS, Rotterdam, 91 pp., 1983.

H. Nguluma: "Case Studies of Production Units of Alternative Binders in Dar es Salaam, Tanzania", Centre for Housing Studies, 1994.

A.A. Nyiti: "Development of the Building Materials Industry (with particular reference to the cement industry) in Tanzania", Ministry of Industry and Trade, Dar es Salaam, 38 pp., undated.

R.D. Reuben: "Amboni Quicklime: Simple Lime Production at Amboni, Tanga District, Tanzania", 1994.

J.K. Rwoga: "A country (Tanzanian) paper presented to the UNIDO/PTA Seminar on Technical Co-operation and mobilization of investment resources for the cement industry in the PTA

Sub-Region", TSC, Dar es Salaam, 10 pp. + map, 1993.

J. Sauni: "Pozzolime as a cement substitute", paper presented to the Seminar on Rural Housing, Arusha, 1981.

J.H. Sakula: "Lime-pozzolana project report January-June 1980", final progress report to SIDO, 14 pp., 1980a.

J.H. Sakula: "Report on feasibility study for lime-pozzolana production in Mbeya region", SIDO, 21 pp. + appendices, 1980b.

J.H. Sakula: "Pozzolime as a cement substitute", paper presented to the Conference: *Towards a National Housing Policy*, Arusha, 5 pp., 1980c.

R. Salomon: "Housing Conditions in Tanzania", paper presented to the Conference: *Towards a National Housing Policy*, Arusha, pp. 12, 1980.

S. Sinha: *Mini-cement: a review of Indian experience*, ITDG, London, 116 pp., 1990.

R. Spence and J. Sakula: "Lime-pozzolana as an alternative cementing material", pp. 96-98 in *Appropriate Technology in Civil Engineering*, ICE, London, 1980.

D.F. Stewart and B. Muhegi: "Strategies for meeting Tanzania's future cement needs" in *National Resources Forum*, pp. 294-302, November 1989.

J.S. Suleiman: "Production of Lime in Zanzibar", STCDA, Zanzibar, 1994.

United Nations Centre for Human Settlements: "United Republic of Tanzania: lime-pozzolana project at Oldonyo Sambu", pp. 42-45 in *The Use of selected indigenous building materials with potential for wide application in developing countries*, Nairobi, 1985.

United Nations Centre for Human Settlements: *Journal of the Network of African Countries on local building materials and technologies*, Vol.1 No.1, April 1989.

United Nations Centre for Human Settlements: "Development of national technological capacity for production of indigenous building materials", Nairobi, 83 pp., 1991a.

United Nations Centre for Human Settlements: Energy for Building, Nairobi, 104 pp., 1991b.

United Nations Centre for Human Settlements: "Building Materials for Housing", pp. 1-19 in the *Journal of the Network of African Countries on local building materials and technologies*, Vol.2 No.2, Nairobi, May 1993.

United Nations Economic Commission for Africa: "Development of the Production of Lime, Pozzolana and Lime-Pozzolana products in the African Region (a preliminary investigation of existing situation and potential), Addis Ababa, 12 pp. + appendices, 1983.

United Nations Industrial Development Organization: "Action Programme to support the dynamic development of the building materials industry in the PTA Sub-region -Proposal for the second and final stage of the project", Vienna, 115 pp., 1992.

United Nations Industrial Development Organization: "Recommendations for action, at the enterprise, national, sub-regional and international level to increase the efficiency of the cement industry in the PTA", UNIDO-PTA Seminar, Kigali, 63 pp., 1993.

Uganda

W. Balu-Tabaaro, W. Okello and B. Kakuru

Uganda has an area of 94000 sq. miles. According to the provisional results of the 1991 Population and Housing Census, its present population is estimated at about 16.5 million with an overall growth rate of 2.5% per annum between 1980 and 1991, and an urban growth rate of 4.8%. The average household size is 5.7. The estimated housing stock is 2,690,900 units and there is a backlog of 235,904 units in the country.

These growth rates are un-matched by the availability of natural resources and the current economic growth rate of 5.0% per annum. The dependency ratio of 54% in the population affects productivity. There is also over-dependency on agriculture which constitutes up to 80% of the country's economic base; 90% of the population are employed in and derive their livelihood from agriculture or the rural sector. Agriculture contributes to over 80% of export earnings and 50% of the GNP.

To manage inflation has proved quite difficult and it has remained at above 20% despite the plan to bring it down to 15% by 1994. Interest rates are high: 18% at the UCB and 22-23% at other Commercial banks. The fixing of these interests at high levels has affected lending and borrowing capacities as well as prices of goods and services.

The above state of the economy has enormously affected all sectors including the manufacturing sector, where both public and private resources need to be invested in the development of infrastructure for the production of various goods like cement and lime. These items require heavy and long term investments. Hima and Tororo Cement Factories (both under UCI) are the sole producers of cement in the country while lime is produced on a small scale by local lime enterprises. Both cement factories are at an advanced stage of privatization. All lime enterprises, which are much smaller, are privately owned, but the government is assisting them in upgrading the quality of lime by developing more fuel efficient and better kilns.

The last years have seen growth in the building and construction industry of 2.9% in 1991, 5.9% in 1992 and an estimated 6% in 1993 according to the Macro-Economic Department of the Ministry of Finance and Economic Planning. With increased economic activity and investment in the economy it is anticipated that further growth may be registered. This will exert heavy demands on construction materials such as cement and lime.

Binders

Cement and lime have for long been the only common conventional binders in the construction sector. Other binders now include pozzolanic materials, bitumen, molasses, etc, but their usage is still limited mainly to pilot projects. A recent National Survey revealed that the national production and consumption potential stood at 300,000 tonnes per annum for cement and 150,000 tonnes per annum for lime respectively. Most of the cement and lime are consumed in building and road construction. Investment in cement and other binders has had a low priority in national budgetary allocations. More recently, though, the government has accorded high priority to the manufacture of building materials amongst which cement and lime are the leading items.

Cement production

The Uganda Cement Industry Limited (UCI) was incorporated in 1952. Its majority ordinary shareholder is the Uganda Development Corporation Limited (UDC) with a direct participation of 51% and a total interest of 86% in the Company. UCI owns both Tororo and Hima cement factories. The Hima factory which was established in 1967 has been treated as an extension of the older Tororo factory in corporate terms so that both factories are treated as subsidiaries of UCI and each is headed by a factory manager.

Tororo Cement Factory is situated in Tororo District some 220 kms from Kampala in Eastern Uganda (see map 3); it started production in 1954 with a commissioned capacity of 150 t/d, later on increasing to 350 t/d. The cement factory was originally constructed to supply cement to the Owen Falls Dam project which was under construction at that time. Current production at Tororo is below 10,000 t/y.

Hima Cement Factory is situated in Kasese district, some 450 kms from Kampala in Western Uganda. Production started there in 1970 with an installed capacity of 300 t/d. This line operated with an output of up to 67% capacity, or 67000 t/y in 1973. The installation of a second production line with a rated capacity of 600 t/d (200,000 t/y) started in 1970. However due to the expulsion of Asians and other foreigners in 1972, the English firm Vickers, which had supplied the machinery, abandoned the job when it was still incomplete. Current production at Hima stands at about 50,000 t/y and is increasing.

Lime production

Equator Lime, Kilembe Lime Works, Tororo Lime,

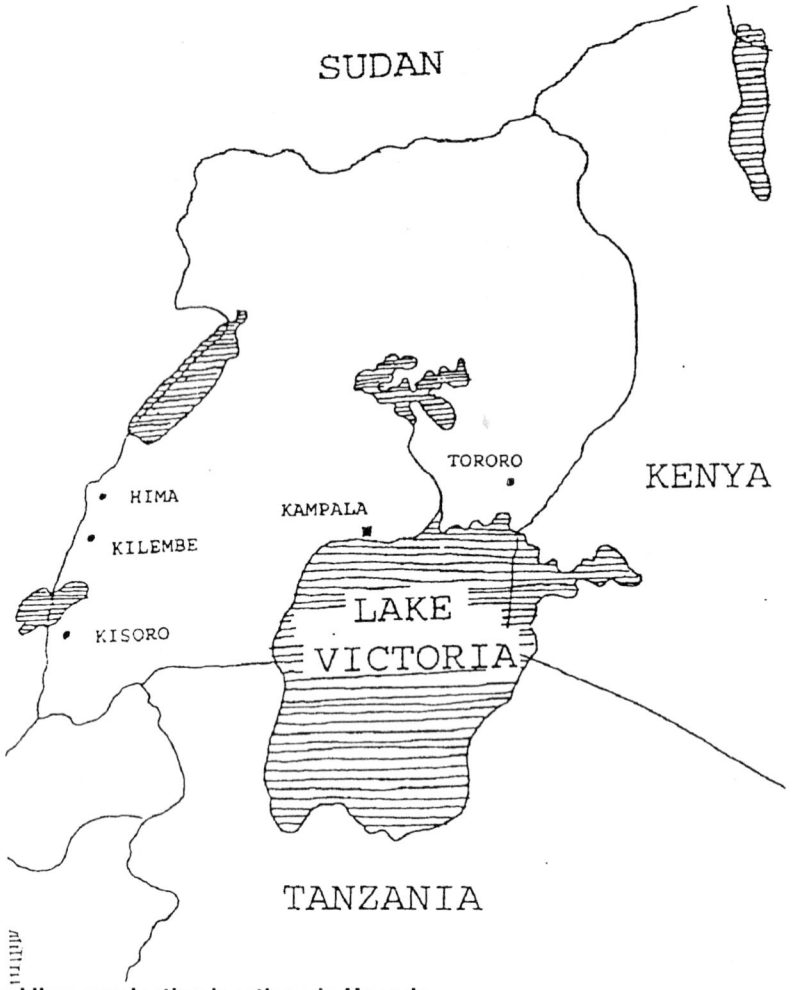

Map 3: Cement and lime production locations in Uganda

Kigezi Twimukye Lime, NEC - Lime Enterprises, Muhokya (and others which constitute less than 5% of the total) are the sole lime producers in the country; they produce mainly lime for roadworks. A substantial quantity of lime is normally imported from neighbouring countries, especially for industrial uses.

Kigezi Twimukye is an indigenous Ugandan-owned and operated company. It has its head office in Kabale Town and its limestone reserves and mines are situated about 15 miles from Kisoro Town on the Kisoro-Busanza Road. It started producing lime in 1976; to date, it has mined 450 tonnes of lime. The company wants to increase and improve the lime they produce but they are hampered by lack of capital and technology to build better kilns and lack of equipment and machinery to mine and prepare the limestone. They have equity in the form of land and buildings. A loan would enable them to secure these missing assets.

Equator Lime get their raw materials from Kaku River limestone reserves. The Equator Lime Works is located approximately 4 km from Kasese on Kasese-Mbarara road. The limestone is a calcareous rock.

The Kilembe Lime Works is located approximately 500 miles from the Hima Cement Factory. The limestone used for lime production is mined from the same deposit used for cement manufacture.

The Tororo Lime Works is situated within 3 km radius from Tororo town. The limestone is of volcanic origin and contains a substantial quantity of phosphate. It is mainly limestone unfit for cement production that is used in the local production of lime.

Raw materials

Hima Cement factory

This plant is located about 15 miles from Kasese Town on the Kasese-Fort Portal road.

The limestone deposit at Hima is of sedimentary origin and believed to have been of chemical deposition. It exists in two layers separated by a water-table which in a number of places is slightly over 3m below the ground level. The total deposit is 5.40 - 9.00 m thick. The top layer which is being worked at the moment consist of layers of hard massive limestone interbedded with semifriable calcaranites and/or marl. It has been estimated that there are about 23 million tonnes of limestone of which 18 million tonnes have been proven. The main block contains about 14 million tonnes suitable as raw material for cement production (see Table 8). The overburden is minimal and rarely reaches 0.30m. Half of the 14 million tonnes lie above the water table and over the years, about 1 million have been used. If Hima were to produce about 400,000 tonnes of cement per year, this limestone would last about 20 years. If the other reserves are also exploited by blending, then the limestone would last about 33 years.

The clay deposit is about 3 km to the west of the factory. It is a reworked sedimentary deposit consisting mainly of clay quartz grains and limestone granules. The distribution of the components is not uniform in both extent and depth, but there is more of the argillaceous components up to a few feet while quartz grains and limestone tend to increase with depth.

Gypsum is added to the clinker to retard the setting of cement. The clinker is mixed with about 5% gypsum and ground to a fine powder in the cement mill. Up to 1990, gypsum was being imported from Kenya. However, with the start of exploitation of the Bundibugyo (Kibuku) deposits, Hima Cement Factory now obtains supplies locally.

Fuel oil is used mainly for firing the hot gas producer for the raw mill and the rotary kiln. Hima Cement Works currently has an arrangement with Total (U) Ltd whereby the latter supplies fuel oil to the factory and is paid after the sales of cement.

Table 8: Analysis of the main raw materials used at Hima Cement Factory

Constituent	Limestone	Clay	Gypsum
LOI	41.570	9.640	10.36
SiO_2	55.050	55.030	11.84
Al_2O_3	0.580	13.960	3.44
Fe_2O_3	0.790	8.760	0.32
TiO_2	0.040	1.040	
CaO	49.780	4.560	31.560
MgO	1.460	1.850	0.62
SO_3	0.420	0.160	40.17
K_2O	0.080	2.330	
Na_2O	0.090	0.880	
Cl	0.003	0.011	
P_2O_5	0.090	1.400	
$S.R_2$	3.690	2.420	
A/F	0.730	1.590	
Purity			86.37

The weekly raw material requirements for the factory when various production targets are considered are given in Table 9.

Table 9: Weekly raw material requirements at Hima for various production scenarios

Production scenario (t/day)	Limestone	Clay	Gypsum	Fuel oil (litres)
600	6,050	1,068	210	420,000[1]
900	9,077	1,602	315	598,000[2]
1,200	12,103	2,136	420	798,000
1,500	15,128	2,670	525	997,500

1. A fuel consumption of 100 litres/tonne clinker assumed.
2. A fuel consumption of 95 litres/tonne clinker assumed for production capacities above 600 t/day

Tororo Cement factory

This plant is situated about 15 km from the centre of Tororo town on the Malaba-Jinja road. Its main raw materials are the Tororo/Sukulu carbonatites. The Tororo Carbonatite Complex consists of Lime kiln Hill, Cave Hill and Tororo Rock. This limestone has a high percentage of ferrous oxide; a typical sample from Sukulu Hills has a total $CaCO_3$ contents of 85 - 90%; the limestone contains also substantial quantities of phosphate.

The clay presently being used in the raw meal mix is being brought from Malaba which is about 13 km from Tororo Cement Plant nearer to the Kenya border. This clay has a high percentage of Fe_2O_3. Good quality clay is found in Mbale, which is at a distance of about 40 km from Tororo Cement Plant. The Mbale clay was inspected and chemically tested; it has high silica, alumina and iron contents and sufficient deposits of clay exist at site. The clay deposit can easily be excavated by establishing a mechanized or a manual quarry.

Lime

Basically all lime producing units are located at or near known limestone reserves. Other unquantified reserves are in Arua and Kisoro districts. The extraction, production processes and technology used have been described in detailed case studies in the next chapter for: Nyalakoti Lime Co-operative at Tororo, Kilembe Lime Works at Kasese, Equator Lime Works at Kasese, and Kigezi Twimukye Lime Works at Kigezi.

Pozzolana

Raw materials for pozzolanic binders or cements are volcanic ash, rice husk ash and lime. For volcanic pozzolanas, these come from Kisoro, Rubanda, and Bunyaruguru volcanic fields. Small amounts of rice husk ash come from the rice growing areas of Iganga, Tororo, and Mbale. Some pozzolanic materials are also obtained from reject clay bricks. And there is the potential to use slag from Jinja, which is available in quantities of around 10,000 t/y.

Production of Binders

Cement

Though total installed capacity of the two cement plants in Uganda is 450,000 tonnes per annum, the actual production now stands at 10-13% of that. The main reasons for this low production include: breakdown in infrastructure due to civil wars; shortage of managerial and technical skills, and shortage of working capital especially foreign exchange for day to day industrial operations. There is an acute shortage of spare parts; maintenance of machinery and equipment has been grossly neglected. All operations need to be properly planned. Workers need vocational training especially the burners, mill operators, electricians and quality control personnel.

The production statistics of Hima Cement Factory from 1970 up to October 1993 are presented in table 11. It is worth noting that production has picked up since 1990 due to rehabilitation of vital production units. The production target for 1993 was put at about 60,000 tonnes but it is likely to be nearer to 50,000 tonnes or 25% of installed capacity for Line II.

Tororo Cement Factory started production in 1953 when kiln I of 150 t/d capacity was commissioned. Kiln II started production in 1957 with a capacity of 200 t/d. Up to now due to old age and lack of spares, production has been erratic with production ranging from 10,000 tonnes to 30,000 tonnes. At present the factory which has been offered for privatization is working on and off. All the cement produced is sold locally mainly in Tororo and Mbale and some of it finds its way to Jinja and Kampala.

From the two factories, there is hardly any cement that is exported to any of Uganda's neighbouring countries. Instead, a lot of cement is imported from Kenya and Tanzania to fill in the shortage in supply. Annual cement demand is close to 200,000 tonnes, and according to one survey possibly as high as 300,000 tonnes.

Table 10: Production Statistics for Tororo Cement Factory (Capacity 120,000 t/y)

Year	Production(t/y)	Capacity Utilization(%)
1986	9,645	8.0
1987	9,482	7.9
1988	5,981	6.0
1989	12,015	10.0
1990	9,683	8.0
1991	10,446	8.7
1992	4,632	3.9

Table 11: Production statistics for Hima Cement Factory (Installed Capacity 100,000 t/y-line 1)

Year	Limestone	Clay	Production	Capacity Utilization (%)
1970	58,226	17,291	44,264	44.5
1971	75,824	20,353	57,462	57.5
1972	57,615	10,076	39,007	39.0
1973	96,882	20,331	66,384	66.4
1974	98,614	15,625	62,817	62.9
1975	64,199	20,623	50,243	50.2
1976	62,956	20,446	48,170	48.2
1977	54,017	9,475	40,140	40.1
1978	47,333	12,190	31,788	31.8
1979	16,453	4,670	12,508	12.5
1981	783	—	506	0.5
1982	17,448	814	7,766	7.8
1983	22,229	2,704	17,298	17.3
1984	16,245	4,131	11,099	11.1
1985	7,329	3,755	6,911	6.9
1986	17,448	814	6,722	6.7
1987	9,600	2,096	6,426	6.4
1988	12,879	1,639	8,985	9.0
1989	6,646	4,551	4,365	4.4
1990	21,998	3,384	17,242	17.2
1991	27,530	4,860	20,062	20.1
1992	48,016	8,473	33,249	33.2
1993	75,831	13,360	47,292	47.3

Lime

Lime is produced at Kilembe Lime Works; Hima Equator Lime; Muhokya, Kaku River, Kisoro, Tororo Lime Works and at other small scale batch plants in Muhokya, Kasese. Production levels in these places are shown in Table 12.

Table 12: Average lime production per annum

Company	Production (t/y)
Tororo Lime Works	8,000 - 10,000
Kilembe Lime Works	8,400
Equator Lime Works	200
Kigezi Twimukye Lime Co.	40
Others	80 - 170
Approximate total	**20,000**

Overall national production is about 20,000 tonnes of lime per year. All the lime produced is consumed by road construction, house construction - as binder and whitewash, some for chicken feed and small amounts for agricultural lime. Most of the lime used in road construction is imported from

Kenya, Italy, etc. Roadworks, building and industrial uses are estimated to need about 150,000 tonnes of lime per annum.

Pozzolana

At present there is little commercial production of lime pozzolanas in Uganda. A few tonnes are being produced by Equator Lime Co. Ltd using lime from the lime works and volcanic ash from the Bunyaruguru volcanic fields. There is also a pilot production unit at the Department of Geological Survey and Mines, Entebbe where 1 tonne per day of pozzolana cement is produced. The lime is obtained mainly from Equator lime and pozzolanas include reject clay bricks from Uganda Clays, Kajjansi, and volcanic ash from Kisoro, Kabale and Bunyaruguru. The cement is produced by grinding the pozzolanas using a 3' x 3' Denver ball mill.

The grinding time ranges from 2 hours for reject bricks to 4 hours for volcanic ash. After this initial grinding, the pozzolana is interground with lime for 1 hour to obtain a homogeneous mixture which is then bagged ready for use/sale. At every end of grinding a sample is taken to check its fineness, using a Blaine Apparatus; another sample is taken of the intermixed pozzolana-lime cement to test the grain size, the compression strength, shrinkage etc as a means of quality control. The unit is in the process of increasing the production capacity to 9-20 tonnes per 24 hours.

Plans are underway to start a 20 tonne per day plant in Kabale town. An industrial plot has already been acquired from the Municipal Council and is being serviced with water and electricity. It is envisaged that the Rubanda pozzolana (volcanic ash) and the Busanza lime (based on Kaku River limestone) will be utilized in this scheme. The total cost of equipment, etc is estimated at US$225,000 including working capital. One bag of 50 kg of this pozzolana cement is estimated to cost USh3,000 ($3).

Because very little pozzolana cement is being produced, the consumption is still small, about 300-500 tonnes per year. Consumption is expected to rise once the Equator Lime Pozzolana Cement factory improves on their fineness and the Volcano (U) Ltd factory comes on stream.

Uses and distribution of the binders

Use of cement and lime

Table 13 below shows the general pattern for distribution and use of cement and lime in selected districts. The figures refer to quantities consumed as registered by the respective Trade and Development Offices and some main dealers prior to the liberalization of trade about two years ago. The actual current figures could not be precisely ascertained because quite a sizeable volume of the binders are now handled by several small scale dealers who do not maintain records partly for fear of taxation.

It is evident that up to 90% of the cement consumed in the country is used in the building and road works, while water supply (e.g., spring protection, piping, etc) and other minor uses account for the remaining 10%. The bulk of the locally produced lime is consumed in road works, with a small amount used for building works such as masonry and rendering of buildings. Lime for chemical and industrial uses is often imported from neighbouring countries and beyond; for example 1000 tonnes of lime is imported annually for sugar refining and leather tanning industries. The importation is mainly prompted by the relatively low quality of the locally produced lime. Earlier attempts to manufacture lime in Tororo for agricultural uses stalled due to lack of resources (finance and technology). Plans are underway to review the programme.

Table 13: Consumption of binders in Uganda

Town	Cement T/month	Cement T/year	Lime T/month	Lime T/year
Kabale	125.5	1506	35	425
Masindi	150	1800	5	60
Kasese	200	2400	100	1,200
Masaka	300	3600	75	900
Jinja	360	4320	120	1,440
Kampala	2,175	26,100	200	2,400
Lira	60	720	—	—
Tororo	60	720	75	900

The above table shows that the centres with the most construction work are Kampala, Jinja and Masaka in order of ranking. Common factors governing the demand for binders include:

- Building construction for individuals, institutions, industries, etc. With improved economic and political stability the country is now witnessing a rapid growth in the construction sector. Currently, the national annual demand for cement is estimated at 200,000 tonnes at least.
- Road construction projects. Over 1,200 km of road length has been earmarked for construction within the next years. These will require approximately 125,000 tonnes of lime.
- Water supply and sanitation projects. These involve spring protection, wells and gravity schemes, bore hole drilling, etc.

The price for Uganda factories is much higher than for Kenya and Tanzania because of the low capacity utilization of the local factories and the fact that these two countries charge a much lower price for exported cement than for the one consumed locally. However, the current prices are much lower than a few years ago. The price reductions are partially forced by the current economic liberalization and need to compete with imports at world market rates. When the transport charges from Kenya and Tanzania are added to the ex-factory prices, Hima Cement competes favourably with the imported cement. In addition, with increased production levels and better capacity utilization, the Hima cement price will fall even further hence making it more competitive.

The transport charges by rail from Hima are about $ 20 per tonne to Kampala and U$ 23 to Jinja.

Table 14: Ex-factory price per tonne of cement in East Africa

Country	Price per tonne ('93)	Import duty	Sales tax
Kenya	65 (Bamburi)	3%	10%
Tanzania	65 (Tanga)	3%	10%
Uganda			
- Tororo	105	10%	10%
- Hima	120	10%	10%

Distribution

Hima markets its cement as Hodari Portland Cement in 50 kg (4ply) paper bags. Due to the current low production in a high demand situation, Hima is able to sell all the cement produced. The customers for Hodari Portland Cement include: government bodies, local contractors, non-governmental organizations, traders who resell the cement on retail basis and international organizations and agencies undertaking building projects. Some of the clients especially NGOs, UNICEF, contractors and traders from towns not served by the railway, purchase the cement directly from the factory or the Kasese depot (300T). This is convenient as double handling is eliminated.

For Kampala and beyond, the cement is delivered by rail to the Kampala depot (1,000T), and Jinja depot (300T). Previously, UCI had depots in Kasese, Fort Portal, Kampala, Jinja, Masaka, Mbarara, Tororo, Mbale, Gulu and Arua. Most depots except the Kampala depot had capacities ranging between 300 and 500T. Currently only those at Kasese, Kampala, Jinja and Tororo are operational. For future marketing especially when production has picked up, UCI intends to have a depot or agent in every major district town. The non-operational depots will be re-opened. In summary the following distribution pattern is anticipated.

Table 15: Anticipated cement distribution pattern in Uganda

	Road %	Rail %	Total %
Local consumption (southern+western)	10		10
Regional customers (central, eastern, northern)		50	50
Major contractors	20		20
Depots	5	15	20
Total	**35**	**65**	**100**

Production technology

Cement production technology at Hima

The factory possesses three drilling rigs; holes used are 4" diameter. About 25-30 holes are charged at a time for each blasting. Due to the ground water level, there is a single bench of about 3 metres height. The drilling pattern is not well planned and documented.

For blasting, each hole is charged with dynamite (Gelatine) as the bottom charge and ANFO as the column charge. A non-electric ignition system is used without any short interval blasting delays between the holes. The drilling and blasting method described above results in the production of many oversized boulders. This leads to time-consuming secondary blasting and hence extra costs. Sometimes the large boulders are fed to the crusher; they block the crusher feed chute resulting in reduction in crusher capacity. As the plant production capacity picks up, more crusher production will be required and full capacity can only be achieved with finer fragmentation of the limestone with appropriate primary blasting.

Three 20 tonne trucks transport all the limestone required to the crusher. For the targeted highest production of 1300 t/d of clinker, the crusher needs to be supplied with 200 t/h of limestone, which would ideally require 5 trucks. Clay is currently transported by the same trucks, since the three available 10 tonne dumpers are all out of use. It is planned to acquire more dumpers.

In the recent past, due to the fact that the clay dryer has been out of action, the clay has been excavated only during the dry season. It has been allowed to dry in a stockpile before transportation to the factory. This procedure is to be recommended, but can only be practised to a limited extent due to the limited transport capacity. For increased production capacities, the clay will be dried and crushed in a clay processing facility to be acquired.

The three major raw materials are crushed separately in a double rotor hammer crusher with an installed capacity of 100 t/h. The crushed raw materials are transported by an inclined belt conveyor for storage in separate stores from which they are extracted for further processing as appropriate.

The raw materials are transferred from the store to separate bins (hoppers) by means of an overhead crane. They are then weighed in a predetermined ratio by the feeders located at the bottom of the bins. The raw mix while being ground in the mill is at the same time dried by passing either waste gases from the kiln or hot gases produced in the auxiliary oil burner.

The raw meal is extracted from the storage silo, accurately weighed and introduced into the kiln via the preheater, a system of vertical arrangements of cyclones down which the gravity flow of the meal is slowed by counter current of hot gases from the kiln drawn by the preheater fan. As the material slowly drops down the preheater, it is gradually heated up to about 800°C before entering the kiln. In the kiln, a tube inclined at about 3°, the material slowly travels by gravity as the kiln rotates and gets gradually heated by hot gases produced by firing fuel oil from the other end. Eventually at about 1450°C, the material partially melts forming black nodules called clinker which are discharged from the kiln, cooled and passed for storage.

The clinker is mixed accurately using weigh feeders, with about 5% gypsum and ground to a fine powder in the mills. The product is then cement which is passed to storage silos to await packaging. Cement is drawn from the silos and packed in 50 kg bags when required.

Production constraints at Hima cement factory include:

- The quarry operations are constrained because of lack of properly functioning large capacity compressors for the drilling rigs and of loading and transport equipment, especially for limestone.

- The current crusher is used for crushing limestone, clay and gypsum. This reduces its production capacity. Moreover, the material which is crushed has a high moisture content especially the limestone fines and clay. This often leads to clogging of the crusher and coating of the liners especially during the wet season. Another problem is that of oversized boulders which block the crusher feeder from time to time. The crusher has to be stopped and a mobile crane used to extract the boulders. All these factors lead to a reduction of crushing capacity to about 70 t/h instead of 100 t/h.

- The rated capacity of 48 t/h of the raw mill has never been achieved mainly due to excessive moisture in the raw materials. Other bottlenecks are the classifier and air slide conveyor; and bucket elevator system. The average output of the raw mill due to the above problems is currently restricted to 25 t/h. There is therefore a need for drying facilities for both clay and limestone. A modification in the auxiliary burner has reduced the volume of gases it can produce. This further lowers the efficiency of the mill when the burner is being used. The blending system needs to be overhauled.

- The electrostatic precipitator is currently not in use and other dust collection systems are not efficient, as a result a lot of raw meal is lost through the chimney. This loss has been estimated to be more than 10% leading to a reduced factory output. It has also adverse implications for the surrounding environment.

- The present preheater fan does not have the capacity to provide the specific volume and pressure of air through the preheater system to achieve maximum kiln output. A higher volume fan is required, of say 4000 m^3/min. Many instruments for controlling and monitoring kiln operations are required and need to be procured and installed.

- The cement packing machines are outdated and inaccurate. Therefore, they lead to underweight cement bags. The bags used for cement packing are about 10 years old and have therefore deteriorated leading to frequent tears during packing and transportation.

- Frequent power interruptions are common, though the situation has improved tremendously due to emergency rehabilitation being carried out on the line by UEB every weekend. Also, Mubuku power station of Kilembe Mines supplies sufficient power to run vital Hima installations when UEB power is off.

- There is a general lack of adequate working capital for the purchase of spare parts and vital components of production machinery which leads to frequent break-downs of various units. The technical staff are currently keeping vital machinery operating through improvisation.

Cement production technology at Tororo

The principal supply of limestone is from Tororo lime kiln quarries, about three kilometres from the factory. The stone is crushed down to 1½" size and brought to the factory by tipper lorries. The lorries tip the stone beneath a travelling crane which conveys it as required to hoppers feeding the raw meal mills. Clay is brought to the factory from Malaba about 13 kms from the works. After passing through a dryer, clay is stored in the clay store.

88% of crushed limestone, 11% of dry clay and 1% of fluorspar are grounded together in the raw mill. The raw meal thus prepared passes to raw meal storage silos. The feed then goes to nodulisers where it is mixed with water and formed into small nodules. These nodules first pass over the preheating grates and then into the kiln itself where reaction takes place at a temperature of about 1400°C. At this temperature the nodules fuse into cement clinker. In this state they pass through a cooler and are carried to the clinker store by an elevator.

The clinker is ground in a mill with 5% of gypsum and the finished product Portland cement passes to cement storage silos. From the storage silos the cement passes to the packing plant where it is filled in 50 kg bags. The final product is despatched from the factory by road and rail transport.

Kiln No. 1 has been running till June 1983 though with constant breakdowns. During this period it was producing some 70-80 tonnes of clinker per day. Many parts of this kiln were deformed and it has been running on a most uneconomical basis. Already in 1983 it was assumed that it would be economically sensible to look for funds to buy a new kiln as an attempt to repair this one would cost almost the same.

Kiln No. II broke down in May 1978, thus reducing the production of cement below its installed capacity. Rehabilitation of Kiln No. II was completed in July 1983 and now again the production figures have touched the lowest level. Breakdowns are excessive and the system as a whole needs rehabilitation. Because of scarcity of funds, the barest minimum expenditure towards rehabilitation is proposed. Kiln II has a pertinent problem of ring formation and eccentric running. Sometimes the ring formation of materials in the burning zone is extending up to 8 m length. So much of the burning materials are stuck in the brick lining that the material flow is blocked and the kiln has to be stopped for cleaning. The high ferrous-oxide content of the raw meal is the basic cause for depositing a hard layer of material on the brick lining throughout the inner circumference. Most of the stoppages of the line can be prevented by properly planning the resources for production schedules and enforcing maintenance management of the highest order which is now missing. While in operation there is a lot of dust in the factory, neither the dust collectors nor the electrostatic precipitator are functioning. Leaking roofs affect storage of raw materials which in itself affects production.

Past production reports have indicated the following production constraints at Tororo cement factory:

• Very often, the star feeder screw shaft breaks down, causing a hold up to the line operation; insufficient inclination of the chute near the disc nodulizer is the main cause of breakage of the screw shafts. The wet material is getting stuck and the remedy to this problem is that discharge from the concrete hoppers should be made by a straight chute to the conveyor and eventually it should be fed to the stock hall and not back to the star feeder screw.

• Water supply to the Tororo Cement Factory has been arranged from three bore holes at different locations. Over the past year at least, only borehole No. 1 has been operating but it was not able to meet the water demand of the factory and the staff houses. Very often cement production was stopped due to a lack of water at the disc nodulizer. This problem could have been solved if someone had taken the initiative to repair or replace the submersible pumps in order to eliminate the costly breakdowns to the line operations.

• The cement factory operations have stopped many times due to non-availability of raw meal. The quarry management and the quarry equipment share responsibility for this down time. Quarry management should reorganize the operations so as to get optimum working hours from the available machinery. Broken down equipment needs spare parts for rehabilitation, the working equipment is old and not very reliable. These need care by planned maintenance while others need replacement.

• Stopping of kiln due to ring formation caused by frequent off-and-on operations. As such, causes of each breakdown should be investigated in detail and remedial solutions should be found.

• Shortage of working capital and especially foreign exchange is a major cause for interruption of operations. A workable solution to overcome the problem of shortage of spare parts, refractories, chemicals and grinding material should be found.

• Because of the kiln running under uncontrolled temperature due to the absence of pyrometers and temperature sensors, the tyre at the outer end of the kiln has been damaged and a length of 4 m of the shell has been deformed near the outlet and it is causing eccentric running of the kiln and frequent falling of bricks.

• Out of the two cranes in the stock hall, only one is working intermittently and is thus causing a great delay in internal transportation of raw meal and clinker. Stock hall roofing too requires urgent attention due to heavy rains sometimes bringing operations to a complete stop.

• Many times line operations have stopped due to non-availability of air supply. Compressed air supply has been arranged as a central system for factory operations. Five compressors are installed in the compressor room, out of which three are beyond economical repairs. Two additional compressors of 600 cfm capacity at 60 psi will be required for smooth operations of the cement plant.

• Too many interruptions to the operation of the kiln are not good. The unsteady operation, of heating and cooling down the kiln too often, decreases the clinker quality, causes coating ring formations in the kiln, harms both brick lining and machinery and as a result decreases production but increases the calorific consumption and thus the cost of production. Every interruption should be followed by a detailed study as to how it could have been prevented. Once a kiln is stopped, to heat it up again will waste another 6000 litres of furnace oil which is an imported item and scarce. It would be worthwhile if furnace oil like diesel and petrol could be procured by government and depots created at Tororo. This will ease the availability of oil.

Lime production

The basic raw material for the production of lime is limestone. Limestone deposits are found in various areas in Uganda, e.g. at Hima and Muhokya in Kasese District, at Tororo in Tororo District, and at Kaku River in Kisoro District.

In order to access the limestone, the overburden is removed by hand in the case of Muhokya, Kisoro and Tororo and by bulldozer at Hima Lime works. In the majority of cases, the material is removed manually with pick axes, spades, hoes. It is only at Hima that drilling using a compressor and blasting are practised. Once the limestone is loosened, it is transported by karais or lorry to the crushing area where it is crushed manually with sledge hammers or by a crusher. The limestone is usually reduced to 80 mm lumps ready for feeding into kilns. The crushed limestone is then loaded into wheelbarrows or wagons and taken to the kiln, where it is loaded manually via a ramp or via rail in the case of Hima lime works.

The kilns used in most cases are batch kilns of heights of up to 3.90m, with the bottom of the kilns lined with steel rods to support the fuel and limestone (see the individual case studies in Section 2). To charge a kiln, alternate layers of fuel (firewood) and limestone are loaded into the kiln until it is full up. Sometimes charcoal is used but this has now become too expensive.

The quicklime is off-loaded into wagons or wheelbarrows by shaking the steel rods at the bottom of the kilns. The loaded wagons or wheelbarrows are pushed manually to the limeshed.

The slaking of quicklime is carried out manually by spraying water on the quicklime. The quicklime crumbles into powder, the hydrated lime. This is then screened manually using primary 6-8 mm sieves and secondary 1-1.5 mm sieves to obtain fine and coarse limes. These are stored ready for bagging. The bagging of lime is done manually. The fine lime which passed through a 1.168 mm sieve is bagged in 20 kg paper bags using a bagging funnel. The coarse lime is stockpiled or bagged into 50 kg plastic bags ready for sale as agricultural lime. The lime is now ready for storage and sale.

Production of Pozzolanic Cements

Raw materials for the production of pozzolanic cements include lime, volcanic ash, diatomite, burnt clay, reject clay bricks, and rice husk ash. At present very little pozzolanic cement is produced, at an experimental plant at Geological Survey and Mines and at Equator Lime Company in Muhokya, Kasese.

In order to produce pozzolanic cement, lime and volcanic ash are mixed in certain proportions. The volcanic ash is first ground to a fineness of at least 3000 cm^2/g (as measured with the Blaine apparatus) in a ball mill. The lime is then mixed or interground with the volcanic ash to create a homogenous mixture. In a batch process, the process is carried out for a certain amount of time while if an air separator/classifier is available, the process can be continuous.

Table 16: Typical production cost of a tonne of lime in Uganda (USh)

Overheads	Required amount	Total cost
Firewood:		
- lorry hire	1 lorry	6,250
- labour	1 lorry	15,000
Raw materials:		
- limestone,	2 tonnes	11,200
- other expenses		20,000
Labour:		
- supervision,	1 person	10,000
- loading of kilns & off loading	5 persons	2,000
- firing 8hrs	2 persons	15,000
Other expenses		5,000
Total costs/tonne		**89,450**

Once the intergrinding is finished, the pozzolana cement is collected and stored and bagged ready for sale. The process though seemingly simple, requires that a high degree of quality control be exercised. This requires that lime of high fineness and quality (65% Ca(OH)$_2$) be used. The volcanic ash must also be reactive (this can be established in the early stages by strength tests and other tests) and continuously tested even when production has started.

Table 17: Estimates of cement consumption in Uganda, 1987-1992 (tonnes)

Year	Domestic production (1)	Imports (2)	Total consumption (3)	(1) as % of (3)
1987	15,904	71,275	87,179	18.2
1988	14,960	111,757	126,717	11.8
1989	17,378	130,655	148,033	11.7
1990	26,920	61,597	88,417	30.4
1991	27,138	102,659	129,797	20.9
1992	37,881	135,879	173,760	21.8

Source: Statistics Department, Ministry of Finance and Economic Planning (1) and Customs Department (2)

Market analysis

Cement

The current demand for cement in Uganda is far beyond what the country can produce. As a result, the country has been forced to import cement so as to satisfy its ever increasing demand, caused by a booming construction industry which has been growing at about 6-10% per annum.

Estimates of cement consumption are given in Table 17 for the years 1987-1992. The domestic consumption was highest for 1992 when a total of 173,760 tonnes were consumed giving a per capita consumption of 10.5 kg. This can be compared with the per capita consumption for the late 1960s and early 1970s which was 22 kg.

There are three major sources of cement supply in Uganda:

- local production at Hima and Tororo factories,
- import from Bamburi in Kenya,
- import from Tanga and Wazo Hill in Tanzania.

Smaller amounts come from donations to various donor-supported projects. It is apparent that Uganda imports most of its cement from Kenya and Tanzania. Imports from both countries have been increasing every year (67% in 1990/1 and 32.4% in 1991/2). The country has been spending huge amounts of foreign exchange on cement importation, for example:

- 3.7 billion Sh in 1990,
- 8.96 billion Sh in 1991,
- 17.2 billion Sh in 1992.

Table 18: Projected demand for cement in Uganda at 6% growth per annum (tonnes)

Year	Actual consumption	Potential demand
1992	173,760	240,000
1993	184,186	254,400
1994	192,237	269,664
1995	206,951	285,844
1996	219,368	302,994
1997	232,530	321,174
1998	246,482	340,444
1999	261,271	360,871
2000	276,947	382,523
2001	293,564	405,475
2002	311,178	429,803
2003	329,848	455,592

The effective demand for cement can be assumed to be higher than the actual consumption because of the low income base of the population, constrained foreign exchange situation in recent times and low domestic cement production. Consumption of cement is further suppressed by a high price due to high production costs. With increased production, the costs will hopefully go down resulting in lower sales price which should stimulate cement consumption. For analysis purposes, the effective demand can be taken to be equivalent to 15 kgs/capita or about 240,000 tonnes/year. This figure is used to project future demand as shown in Table 18. The major assumption is that the construction industry will continue to grow at a rate of 6% per annum (as stated in the Way Forward I and II: Medium Term Economic and Sectoral Strategies, 1991-95: Ministry of Finance and Economic Planning).

The forecast may even be on the conservative side when the rate of rehabilitation and ongoing or planned new construction are considered.

Table 19, below, provides the annual estimate of cement which may be required for housing in the near future, which amounts to about 1.3 million t/y.

Table 19: Cement requirements for shelter (annual estimates for the plan period 1992-2000)

Type of house	Number of units	Cement required (50 kg bags)
a) Up-grading		
- up-grading traditional	17,318	346,300
- semi-permanent	12,370	618,500
- low-cost	10,061	2,012,200
- medium-cost	2,639	791,700
b) New housing (owner-occupied/rental)		
- Low-cost	33,606	10,081,800
- Medium-cost	12,709	6,354,500
- High-cost	6,232	6,232,000
TOTAL	**94,935**	**26,437,060**

Source: National Shelter Strategy for Uganda, Ministry of Housing and Urban Development

Future trends and development issues

Cement

The estimated annual demand for cement in Uganda ranges from approximately 200,000 tonnes to 400,000 tonnes (1989-2000 figures). The demand for cement has been increasing at about 6 - 10% per annum. Apart from the housing sector mentioned before, major consumers of cement will be the following projects, which have a large demand for construction: agro-processing, mining and quarrying, tourism, banking, warehousing, business centres, building construction, manufacturing and assembly.

Most of these projects will be located in (in order of greatest consumption): Kampala, Jinja, Mbarara, Masaka, Mbale and Kasese

In order to meet this demand, apart from importing cement from Kenya and Tanzania, plans are underway to rehabilitate and expand production at both Hima and Tororo Cement Factories. There is even a proposal to start a new factory at Bududa, Mbale, based on the limestone reserves there. (Though a drilling programme was carried out and the indicated reserves were considerable, production was shelved due to high costs that would be incurred in heavy overburden removal).

As for the expansion at Hima and Tororo, this is expected to be effected by new private owners who are likely to take over the factories soon. At the time of writing the report, these two factories had already been put up for privatization.

Lime

Apart from Ordinary Portland Cement, some lime has been produced, albeit on a far smaller scale, at Tororo, Hima, Muhokya and Kaku river in Kisoro. Most of this lime is used in road construction, as a binder and whitewash for mainly low cost houses in the vicinities where this lime is produced. Some lime from Kaku River, Kisoro finds its way into Rwanda and Zaire. The lime produced in all these places has been of low quality and in small quantities due to poor technologies, mainly poor kiln design and construction, compounded with a lack of financial resources. Attempts are being made to put up bigger and more fuel efficient lime kilns to cater for the increased demand from various users: road construction, sugar industries, binders, whitewash and lime-pozzolana cements. The latter is an innovation based on recent research on the use of pozzolanas for the manufacture of pozzolana cements.

Pozzolanas

Pilot production of lime-pozzolana has started at Equator Lime in Muhokya using their improved lime. The quantities are still small and there is a need to improve on their fineness. Another pilot plant is being set up at Kabale based on Kisoro volcanic ash and lime from Kaku River limestones. This plant will be producing a cement, estimated to cost about Sh 3000 per 50 kg bag ($ 3.0). Proposals are at an advanced stage to put up another 30 t/d plant in Kabale industrial area by a local Company, Volcano (U) Ltd, which will also use Kisoro volcanic ash and Busanza lime.

Studies were carried out by the department of Geological Survey and Mines on the possible production of Portland Pozzolana cement by the Hima Factory in Kasese. The results were positive and the company is in the process of sourcing finances to start commercial production of this type of cement.

Environment

There are no clear policies in place to encourage more environmentally friendly exploitation of the natural resources required for binder production, such as the reclamation of quarries, better designed mining and processing methods, control of air pollution, or avoidance of haphazard disposal of wastes/ tailings from the quarries. Large scale harvesting of firewood may also lead to deforestation and soil erosion and consequent drought and famine. The working environment of factories leaves a lot to be desired: no protective gear is provided and this leads to a high labour turn-over.

Energy

There are significant opportunities for energy efficiency improvements in the cement and lime production sectors. In lime production, charcoal is used particularly at Kigezi Twimukye Lime. There are signs of incomplete combustion, fuel wastage and poor quality of product. Energy efficiency initiatives involve redesigning of kilns and efficient operation. Batch kilns require energy to start them up to the required temperature every time they are started. Conversion of these batch kilns to continuously operated kilns may give efficiency gains and use firewood more rationally. At present a prototype continuously fired kiln of 5 t/d capacity is under construction in Tororo with funding by the Economic Commission for Africa (ECA). This kiln will act as a demonstration unit where aspiring lime-producers can be trained in appropriate lime production. Energy conservation issues have a direct bearing on environmental aspects of deforestation and degradation.

Government policy

The goal of the Government's industrialization policy is to establish a strong, viable, sustainable and internationally competitive industrial sector, integrated with the rest of the economy through backward and forward linkages, as an important contributor to the development of an 'integrated, self sustaining national economy'. The general strategy is to reduce the Government's role in the economy. To make best use of its financial resources, the Government is pursuing a three-tier strategy:

- to liquidate those enterprises which are clearly no longer viable;
- to divest those enterprises which would operate much more efficiently in the private sector;
- to restructure those enterprises remaining in the public sector to improve their efficiency and to reduce their reliance and dependence on Government's financial support. In this respect the Government has gone ahead to sell off Hima and Tororo factories.

In order to enhance industrial development the Government encourages research and development for transfer of technology and development of indigenous technology.

Availability and reliability of supportive infrastructure is a prerequisite of industrial development. The Government will endeavour to expand, improve, and increase the supply of power and water, communication facilities, transport and road networks. The rehabilitation of Kampala-Kasese railway line is one of the major priorities to facilitate transportation of products from Kasese.

A long term industrialization policy has to ensure an ecologically sustainable industrial development. Air pollution around Hima factory needs to be addressed by rectifying faults at the electrostatic precipitators.

The Government believes that there is no efficient and effective way of long-term protection of local industries against competition of lower cost and better quality imports. However, it is recognised that unfair competition, e.g by dumping of foreign export subsidies, has to be discouraged (by use of relevant tariff arrangements).

A facilitating banking sector as well as functioning money and capital markets are critical not only to the privatization but also to the overall success the government's industrialization plan. In addressing these, the Government has already liberalized the foreign exchange market and the interest rates, moving to a market-based system of determining key lending and deposit rates by linking them to the auction determined treasury bill rate and by decontrolling all other rates. The Government will therefore continue to use appropriate monetary policy measures to stabilize the value of the Uganda shilling.

Conclusion and recommendations

From the national survey carried out, the demand for binders, especially Ordinary Portland cement is very high. The biggest demand is in the construction/building industry. At the same time, the cost of this cement is very high for the ordinary citizen. The main reasons for the high cost of cement is low capacity utilization at the two cement factories. Because of past political problems, these factories were mismanaged. The cement is also consumed in places far away from production centres and transport costs are high. The cheapest transport system that could have been used, the railway, is inefficient due to inadequate maintenance. It is hoped that if the privatization of the existing factories can be carried out successfully, it can lead to higher capacity utilization and lower production costs that would reduce cement costs to the consumer.

With improved kiln designs, better and more lime can be produced. This would save foreign exchange that is presently spend on importation. It would also supply cheaper lime for use in the construction and building industries. Better quality lime would also enhance lime pozzolana cement production. This would provide an alternative to expensive Portland cement.

It was noted that the current production plants and facilities are obsolete, they lack spare parts and there are frequent break downs which affect productivity. There is therefore a need to overhaul the system and if viable, adopt new technologies. There is also need to look into the use of small scale production technologies that have low capital inputs. Investment into research and development on innovative technologies should be encouraged.

The main sources of limestone for cement and lime are located in Tororo, Hima and Muhokya. There is need to carry out further exploration and evaluation of these and other resources (e.g Bududa limestone, Karamoja limestone/marble) in order to access their utilization in the production of binders. Rational and sympathetic exploitation methods and processes should be encouraged. This will be necessary to avoid misuse of limited resources.

One of the main contributors to the high cost of cement and lime is their distribution costs. It is vital that production be decentralized in order to minimize transport costs. Depots should be established in different locations to enable easier access. There is need to improve the rail, road and water transport systems to enable easier and cheaper distribution systems. In order to sustain development of the binder industry, the mobilisation of local resources should be encouraged. This will reduce overdependence on foreign capital that usually comes with a lot of unachievable preconditions.

It was noted that there is a shortage of well trained management and technical personnel. The lack of this category of workers constrains efficient operation of the binder industries. There is need to invest in training of all levels in order to enhance efficient binder production which in turn will reduce prices.

Energy is the main cost item in the production of cement or lime. These costs eventually raise the price of these binders. In the case of lime, a lot of trees are used as fuel. Not only does the cutting of trees cause environmental degradation but the cost of firewood is rising daily due to scarcity and the cost of transport. It is recommended that fuel-efficient technologies be adopted and research into alternative energy sources be undertaken. In the case of lime producers, they should be encouraged to plant fast-growing trees.

It has been noted that exploitation of resources, e.g. quarrying, tree harvesting, waste disposal, to produce binders entails degradation of the environment. There is a need, therefore, to adopt stringent environmental protection policies. These would include quarry reclamation, air/dust pollution control, better tailings/waste disposal systems and environmentally friendly technologies that can be communaly controlled.

Investment in R&D into better and innovative technologies cannot be over-emphasized. There is a need for both the government and private sector including the AT organizations to pool resources to invest in better and more efficient technologies to enhance low-cost binders.

KENYA

T.J. Anyamba

The provision of affordable housing to all citizens is one of the Kenya government's goals in addition to eradicating poverty, hunger and disease. This goal has however remained elusive, primarily because the types of building materials in use are too expensive for low-income earners. The building by-laws and other statutory legislation have not encouraged the use of alternatives to ordinary Portland cement. In any case, the debate on what constitutes permanent buildings needs to be revisited.

Factory produced building materials will continue to be unaffordable to most Kenyans, which calls for the use of cheaper and non-factory produced binders. Industrially processed binders have an added transport cost to them, as most consumers cannot be within easy reach of the industries. The current minimum wage stands at KSh 1,500 per month; the cost of a 50 kg bag of cement is approximately 40% of this minimum wage. Within this context, it is imperative that owners and builders get the necessary exposure to alternative binders. But locally processed binders which are produced at a low or no cost using indigenous or traditional skills, such as special earths, potash, ash and cowdung, have yet to find use in formal sector housing.

The manufacture of alternative binders to cement will stimulate economic growth in the informal sector, thus creating job opportunities and other income generating activities.

Industrially processed binders have a very high level of energy consumption i.e. electric power, fuel oil and coal. Intervention measures have to be instituted that will help alleviate these high levels of energy consumption. Besides, the transport they require is another big consumer of energy. The production process of most mineral based binders leads to environmental degradation. Besides needing less energy and equipment, small-scale production is also more environmentally friendly.

This study is aimed at assessing the current state of binder production and use in Kenya and at recommending strategies to make binders more affordable and to spread the technology to most low-income earners. There exists adequate information concerning the production and consumption of ordinary Portland cement in this country. However, there is a dearth of literature on other binders currently in use. Social prejudices against the use of certain materials are quite common. The main objective of this study is to promote the use of local materials including alternative binders. These materials are not just a poor man's choice, they have other advantages, e.g. they require less energy in their processing, they rely on low-cost technologies in their application and tend to be more environmentally friendly.

Raw materials

Cement

Cement is a finely ground hydraulic binder used for the preparation of mortar and concrete. The main raw materials for the manufacture of cement are limestone, iron ore, gypsum and shale.

The limestone should be rich in calcium carbonate, over 78%. Limestone provides lime or calcium oxide, CaO, which forms the main oxide in cement. The iron ore provides ferric oxide which apart from controlling the chemistry of the raw meal, provides flux for easy burning. Gypsum is used in the manufacture of cement to provide the sulphur trioxide which prevents early setting of the cement when mixed with water. Shale is the type of clay that provides silica, alumina and iron oxide. Limestones with a high concentration of phosphates are unsuitable for cement production. In addition, certain impurities are bad for cement production even in relatively small quantities.

There are large quantities of these materials in many parts of the country. Over 20 million tonnes of coral limestone can be found in the vicinity of Bamburi Portland Cement Company (BPCC) near Mombasa, together with large reserves of gypsum and shale. Limestone reserves of over 27 million tonnes are found around Athi River near Nairobi and over 10 million tonnes of Kunkur (lime, alumina, silica and iron oxide) reserves in Kajiado. In addition to these deposits, large quantities of raw materials for the manufacture of cement are available in Koru, along the coast South of Mombasa and many parts of Kerio Valley.

These deposits are of high quality and have the capacity of producing high quality products. For example, most of the coastal coral stone has in excess of 83% $CaCO_3$ (calcium carbonate), the gypsum found in this country has approximately 80% $CaSO_4.2H_2O$ (hydrated calcium sulphate) while iron ore has 40-70% Fe_2O_3 (ferric oxide). Although many deposits of limestone are known, only a few have proved to contain sufficient amounts of the quality of raw materials required for full scale production of Portland cement. However, abundant deposits of high quality limestone with low content of magnesia are found in El-wak, Namanga and Kibini. Extensive outcrops of low quality limestone exist also in Voi and South Yatta. At the current rate of extraction, the deposits of coral stone at Bamburi will take approximately 10 years to be depleted, while those of limestone at Sultan Hamud near Athi River will take in excess of 50 years.

The life of the current deposits near Bamburi Portland Cement Company (BPCC) can be stretched to 20 years if a high carbonate sweetener is brought in from other deposits, north or south of Mombasa where ample reserves of coral limestone for continuation of the cement production have been identified. This will allow the use of more marginal quality material from the existing quarry. Other options for BPCC are:

- to open up a new quarry several kilometres from the existing plant, which will increase the production costs.
- to re-locate the plant to the coast south of Mombasa when the deposits in the present quarry are exhausted.

In either case the location of the plant is close to a recreational and tourist beach area, which may lead to restrictions for future expansion and eventually the closure of the existing plant.

The main producers that use these mineral deposits are the BPCC at Mombasa and the East African Portland Cement Company (EAPCC) at Athi River. For the Bamburi plant, coral stone and shale are within 5km radius. Iron ore is found in Homa Bay, 800km away from the plant, and at Jaribuni in Kilifi District. Gypsum is mined 450km away at Kajiado and at Rhoka in Kilifi District (50 km away from Mombasa). The Athi River plant gets its limestone from Sultan Hamud which is approximately 110 km away. Gypsum is mined 75km away at Isenya while iron ore is mined 350km away from the plant, at Homa Bay.

The roads to the quarries within a 5km radius of the Bamburi plant are good earth roads maintained by BPCC. The road to Homa Bay is a good tarmac road except the last 40km which is a poor earth road. Gypsum, which is mined in Kajiado, is accessed by a good tarmac road except the last 14km which is an earth road impassable during the rainy season. The roads to the quarries of the Athi river plant are generally all weather roads. Infrastructural services (water, sewer, power, telephone) are to be found within 14-20km of the Bamburi quarries. The gypsum quarry at Isenya gets its water supply from a borehole.

Lime

Lime is manufactured from limestone. Large deposits of limestone exist all over the country, including Kitui, Wajir, Taita Hills, Makindu, Bamburi, Tiwi, Kajiado, Koru and some parts of Kerio. As discussed earlier, these deposits are capable of producing high quality products. In general they have in excess of 80% calcium carbonate ($CaCO_3$). The estimated reserves of the Koru deposits are in excess of 65 million tonnes; these deposits are being extracted by the Homa Lime Company at the rate of 30,000 t/y. At this rate of extraction, the deposits have an expected life time in excess of two thousand years.

There are various types of limestone deposits to be found in Kenya. These are crystalline limestone, Jurassic to Pleistocene marine limestones, tertiary and lacustine limestones and carbonated limestones. Crystalline limestone as well as Tertiary and Lacustine limestones occur in many parts of the country with the exception of the Western and Nyanza regions. Marine limestones are found in the coast region, north and south of Mombasa, and in the northern part of the Northeastern Region. Carbonatic limestones are found mainly in the Nyanza Region.

Crystalline limestone is a hard limestone with a high content of calcium carbonate, and also varying contents of magnesia. Large deposits, approximately 13 million tonnes, have been identified in the Kibini Hill area near Sultan Hamud, with the quarry at Bissel near Namanga having deposits estimated at 2 million tonnes. It can also to be found at Ortum and Sebit, northeast of Kitale, and at Koru and Songhor.

Marine limestone is a soft coral-type limestone mostly with a high content of calcium carbonate, but mixed with layers of high silica material. It occurs in large deposits along the coast, north and south of Mombasa and in the Northeastern Region.

Tertiary and Lacustine limestones exist both in the Rift Valley and in the Eastern regions. There are also occurrences in the southern part near Magadi, southwest of Nairobi. Carbonated limestones occur in the Western part of Kenya, in the Nyanza Region near Lake Victoria, on the Ugandan border, and in Koru and Songhor areas. The Koru-Songhor limestones are of carbonated volcanic origin extruded into either Mozambican granitoid and biotite gneiss, or Tinderet nephelinite volcanic. The main producers that use these supplies are Kenya Carboxide Ltd, Homa lime Company, Kenya Calcium products, Lime and Cement Company (EA), and Bharat Lime Company in addition to the two cement manufacturing companies earlier mentioned. Most of the limestone quarries are a short distance from the lime plants.

Special earths

Two types of special earths deposits exist in Kenya. These are brown earths and white earths (kaolin). The brown earths are used in the manufacture of bricks, tiles and pipes, while the white earths have traditionally been used as plasters on houses built with mud. Large deposits of brown earths are found in Western, Central and Eastern provinces. White earths exist in most parts of the country. However

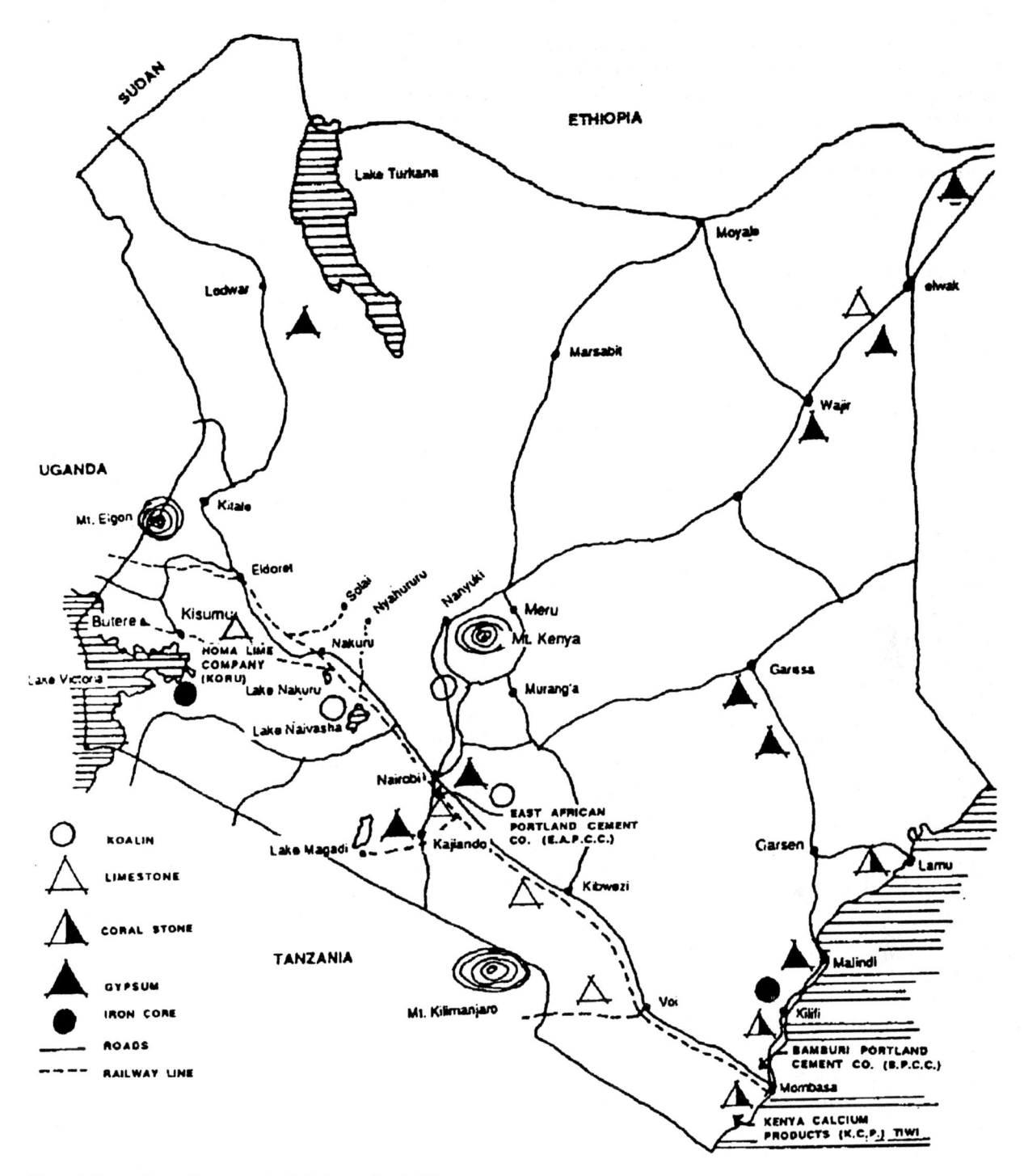

Map 4: Location of raw material deposits in Kenya

high quality kaolin is found in Machakos (Kitandani Hill and Kutaa Hill), Eburu and Olkaria near Naivasha. The Eburu kaolin deposits are over 60,000 tonnes while those at Kitandani and Kutaa hills are approximately 95,000 m³. Hard varieties of laterite and termite hills are found mostly in savannah zones of Kenya. These special earths are used as stabilizers for sandy soils.

Cowdung/ash

Cowdung is used as a wall and floor finish in most mud constructed houses in the country. It is used in its raw form either independently or as a stabilizer in clay soils. The availability of cowdung depends on the number of cows one may have. When decomposed, it produces high quality manure which is useful for organic farming. In some instances, dried cowdung is also used as a fuel. These alternative uses of cowdung minimize its use as a binder.

Ash as a stabilizer to cowdung is found all over the country, with most of the ash resulting from burning of firewood while cooking food. There is no known production of ash specifically for building purposes. The main raw material for the production of ash is timber. However due to population growth, which requires land for settlement, and general deforestation, timber for ash is not readily available.

Pozzolanas

A pozzolana is a silicious and/or aluminous material which will, in the presence of water, combine with lime to form cementitious compounds. It invariably contains silica or alumina in reactive form. Pozzolanas can either be natural or artificial. Examples of natural pozzolanas include volcanic ash, pumice and tuffs and diatomite. Artificial pozzolanas include fired clays, clay reject products, fly ash (pulverized fuel-ash) and agro-waste products such as rice husk ash (treated separately further on), coffee husk ash and bagasse ash. Granulated blast furnace slag is also an artificial pozzolana.

Natural pozzolanas are of volcanic origin and are formed either as violent projections of molten magma into the atmosphere forming a glassy material, or as less violent eruptions forming volcanic ash. Although the chemical composition of the above two products is similar, the glassy material is more reactive with lime than volcanic ash. Pozzolanic materials found in dry or arid regions, are less likely to have been altered chemically by weathering effects and are more likely to retain their original reactivity. Natural pozzolanas can be found in the following areas in Kenya: Londiani, Athi River, Tinderet, Uasin Gishu and Mt. Elgon.

Most pozzolanic materials resulting as by-products of agricultural or industrial processes, are normally produced in large quantities, and constitute a waste problem if they remain unused. Pozzolanas are not cementitious themselves, but when finely ground and mixed with lime, the mixture will set and harden at ordinary temperatures in the presence of water, just like cement.

The Londiani volcanic materials are reactive; they consist of lavas and formations of welded tuffs and unconsolidated ash. They are younger than the Tinderet and the Koru formations. They offer more promise in respect of pozzolanic properties, as the tuff and ash will have a glassy structure and will be relatively lightly weathered. Chemically they contain more silica than the volcanic materials associated with Tinderet. A plant for the production of a lime-pozzolana binder can easily be set up in the Londiani area. This plant can rely on Koru for the supply of lime and on Londiani itself, for the supply of natural pozzolanas. In addition it be supplied with artificial pozzolanas in the form of rice husks from Ahero and West Kano and bagasse from the Chemelil/Muhoroni sugar belt.

Masonry cement

A newly emerging alternative binder whose main raw materials are cement and various blenders derived from some minerals, including diatomite, tuff, kaolin, etc. There are large deposits of diatomite at Gicheru in Ndeiya-Limuru, containing approximately 800,000 tonnes. There are also large deposits of kaolin at Karatina-Nyeri, used for the manufacture of kenite. Diatomite is also found at Kariandusi near Gilgil. The main producers who use these supplies are the Halai group in Nairobi, Nova Industries Ltd in Athi River and Kariandusi Diatomite Company.

The quarries of these deposits are within easy reach of the production plants and the roads are company maintained. However the road to the Gicheru diatomite quarry is very bad. Infrastructural services are not within easy reach of most quarries. They rely on water supply from boreholes, and power supply from generators.

Rice husk ash

This material is found in the rice producing areas of Mwea, Ahero, West-Kano, and Bunyala. At the moment the production of rice husk is over 8,000 t/y. It is however projected that with the introduction of rice farming in the Tana delta, the amount of rice husk will increase to 27,000 t/y. The combustion of agricultural residues removes the organic matter and produces, in most cases, a silica-rich ash. Of all the common agricultural wastes, rice husks yield the largest quantity of ash, about 20% by weight with the highest silica content, around 93% by weight. It is this high silica content that gives the rice husk ash its pozzolanic properties. It should be noted that the rice husk ash in itself is not a binder, it needs to be activated in the presence of water by lime or possibly cement, in proportions of 1 lime or cement: 2 or 3 rice husk ash to make a binder. The fuel value of rice husks is 14MJ/kg, which means that one tonne of rice husks is equivalent to about half a tonne of coal or nearly half a tonne of fuel oil. The road network in most of these rice producing areas is all-weather, water supply is readily available.

Gypsum

Gypsum is available either as natural gypsum or as an industrial by-product. In composition, gypsum is hydrated calcium sulphate ($CaSO_4.2H_2O$) and is found in large quantities near Garissa, Malindi and between Athi River and Kajiado. The deposits of gypsum at Roka Malindi are in excess of 150,000 tonnes while those at El Wak are approximately 2 billion tonnes. Most of the deposits in Kajiado area are exploited by the East African Portland Cement Company, while those at El Wak remain unexploited. Like the other raw materials, infrastructural services are not always available, and have to be improvised.

Binder Production

Cement

There are two main cement producing companies in Kenya, namely: Bamburi Portland Cement Company (BPCC) located at Bamburi near Mombasa and East African Portland Cement Company (EAPCC) at Athi River near Nairobi. The BPCC plant produces about 1.1 million tonnes of cement per annum, while the EAPCC plant produces about 330,000 t/y. This gives a total annual production of approximately 1.43 million t/y. Their installed capacities are 1.2 million t/y and 350,000 t/y respectively. Table 20 shows annual production levels of cement between 1980 and 1994.

The BPCC plant is a dry process plant with 6 shaft kilns and 2 rotary kilns with 4-stage cyclone preheaters and back end firing. This plant started operation in 1954 and creates employment for about 800 people. Productivity at the plant currently stands at 1.76 mh/t which is by far the best in the region.

The plant is owned by the Kenya Government, Kenyan public and overseas investors in the following share division:

Kenya Government 15.8%
Kenyan public 11.0%
Overseas investors 73.2%

The overseas investors are Blue Circle of the United Kingdom and Cementia/Lafarge Coppee of Switzerland.

The basic energy types used at BPCC are electricity, heavy fuel oil, coal and industrial fuel oil. The fuel oil and coal are both imported and are consumed at the rate of 18,000 and 115,000 t/y respectively (approximately $4,131 \times 10^6$ MJ per annum). Currently the cost per tonne of cement is KSh 6094 ($110.8). With an energy consumption of approximately 4,000 MJ/t of cement produced. Although the plant has been written off (no book value), it is well maintained and still efficient. There exist some production constraints and problems, such as:

- Old plant with some worn out or obsolete equipment
- Limited reserves in the existing quarry and thus a need for opening of a new quarry at several kilometers from the existing plant.
- Location of plant close to recreational and tourist beach area, which may lead to restrictions for future expansion and eventually to closure of the existing plant near Mombasa.

After the removal of foreign exchange controls, BPCC now enjoys automatic retention of foreign currency earned through exports. The Kenya Government has also relaxed the rules on the payment of the dividends to overseas shareholders. This re-

Table 20: Cement production at EAPCC (Athi River) and BPCC (Bamburi)

Plant year	EAPCC '000 t	% Capital utilization	% Share	BPCC '000 t	% Capital utilization	% Share	Total
1980	293.0	83.7	22.6	1,005.0	74.4	83.8	1,298
1981	270.0	77.1	20.3	1,060.2	79.7	88.4	1,330
1982	300.0	85.7	21.7	1,082.0	78.3	90.2[b]	1,382
1983	287.0	82.0	22.5	987.0	77.5	82.3	1,274
1984	320.3	91.5	27.1	864.0	72.9	72.0	1,184
1985	317.5	90.7	27.2	847.3	72.8	70.6	1,165
1986	338.8	96.8	28.3[c]	860.0	71.7	71.8	1,199
1987	325.5	93.0	26.2	918.5	73.8	76.6	1,244
1988	322.2	92.1	26.8	878.8	73.2	73.2	1,201
1989	320.4	91.5	24.3	996.0	75.7	83.0	1,316
1990	332.2	94.99	23.0	1,111.0	77.0	92.6	1,443
1991	304.6	87.0	20.9	1,151.0	79.1	95.9	1,456
1992	323.5	92.4	22.4	1,125.0	77.6	93.8	1,449
1993	279.7	79.9	19.6	1,150.8	80.4	85.9	1,431
1994	339.3[a]	96.9	21.0	1,285.7[a]	79.0	107.1	1,626

Notes: a. Highest annual production above the installed capacity in 1994 due to the production of Portland Pozzolana cement
b. Highest market share for BPCC
c. Highest market share for EAPCC

Source: UNIDO 1992 and own calculations based on data from CBS and EAPCC

laxation of the rules is an incentive for attracting foreign investment in Kenya.

The main future plan at BPCC is to increase output by the inclusion of up to 20% pozzolanas. No major production increase is likely as the plant has already pushed its production capacity to its limits. Table 21 shows the production costs breakdown at BPCC.

Table 21: Production costs breakdown at BPCC for bagged cement (per tonne in KSh)

Item	%age	Cost
Coal and oil	18.2	1,109.11
Electricity	11.3	688.62
Fluorspar	0.3	18.28
Iron ore	3.3	201.10
Gypsum	4.5	274.23
Pozzolana	2.9	176.73
Transport for raw materials	1.4	85.32
Paper bags	9.6	585.02
Salaries, wages & social costs	9.0	548.46
Materials issues (spares, etc)	21.7	1,322.40
Administration	10.4	633.78
Finance charges	0.9	54.85
Depreciation	6.5	396.11
TOTAL	**100**	**6,094 ($110.8)**

Sources: 1992 UNIDO study/own computations based on current cement costs

From this table, it can be seen that spare parts, coal, oil and electricity account for more than half of the cost of producing a tonne of cement. For the price of cement to be lowered meaningfully, a saving has to be made on these particular items.

The EAPCC Athi River plant began its operation in 1958, and currently creates employment for approximately 578 people; this translates into a productivity of around 3.5 mh/t. It is a wet process plant with 2 rotary kilns with chain systems and planetary coolers. The ownership of this plant is as follows:

- Kenya Government: 52% (from which 27% is National Social Security Fund-NSSF)
- Public shareholding: 20%
- Overseas investors: 28% (Blue Circle, UK - 14%, Cementia/ Lafarge Coppee, Switzerland - 14%).

This plant uses electricity and fuel oil, imported via Mombasa. Fuel oil consumption is approximately 45,000 t/y, while electricity is consumed at the rate of 2.2 million units per month. The cost of producing one tonne of cement currently stands at KSh 7,846. Heavy costs of transportation of limestone from quarries at about 100km from the cement plant is one of the main production constraints. Because of the wet process, the plant has a high fuel consumption. Lack of imported spare parts and limited financial resources for investment in conversion to the dry process and expansion of the production capacity, are some of the other problems.

Table 22: Production costs breakdown at EAPCC for bagged cement (per tonne in KSh)

Item	%age	Cost '92	Cost '94
Hired transport	7.7	200.0	604.14
Clay - 1%	0.34	8.72	26.68
Electricity	5.7	147.84	447.22
Paper bags	7.2	184.43	564.91
Cement transport	10.2	262.14	800.29
Kiln oil	29.38	756.00	2,305.15
Iron ore	0.84	21.60	65.91
Fixed costs	37.18	956.84	2,917.14
Total	**100**	**2,575.07 ($80.47)**	**7,846 ($142.65)**

Sources: 1992 UNIDO study/own computation based on current cement costs
Note: 1992 $ mean rate: KSh 32
1994 $ mean rate: KSh 55

Future plans include investigating further expansion possibilities based on limestone deposits in the vicinity of present quarries. The company is currently installing a new production line (based on dry process technology) of 1,600 t/d per day. This will increase the production capacity from 350,000 t/y to 540,000 t/y. In addition to this, plans are underway to produce cements with secondary constituents such as slags, pozzolana, fly ash ,etc., which have special properties. This will be in line with global practice which aims at:

- Conservation of key raw materials
- Conservation of energy
- Conservation of the environment by reduced production of carbon dioxide emission per tonne of cement produced

Reduction of potential cement shortages and the consequent need to import cement and imminent loss of foreign exchange.

Lime

The burning or calcination of limestone accomplishes three things:

- The water in the stone is evaporated.
- The limestone is heated to the requisite temperature for chemical dissociation.
- The carbon dioxide is driven off as a gas, leaving the oxides of calcium and magnesium.

The two lime producing companies in Kenya are

the Homa Lime Company, with a factory located at Koru some 80 km east of Kisumu town, and Kenya Calcium Products, a sister company of Homa Lime, located at Tiwi near Mombasa. The Kenya Calcium Products (KCP) produces white hydrate of lime while the Koru plant produces grey hydrate. Tables 23 and 24 provide the figures for annual production of lime between 1981 and 1993 at these locations. Currently the Homa Lime plant at Koru is operating at approximately 30% of its capacity.

There are several small-scale producers of lime who use non-industrial techniques. In Lamu and Wajir there are many small kilns which are used for the production of lime.

Table 23: Annual lime production in Kenya, 1981-93 (tonnes/year)

Year	Production
1981	33,063
1982	26,646
1983	26,166
1984	20,855
1985	13,105
1986	35,000
1987	37,460
1988	28,167
1989	32,167
1990	35,733
1991	12,047
1992	8,771
1993	11,554

Sources: Statistical Abstract 1991 and 1994 figures from Central Bureau of Statistics

The Homa Lime Company plant at Koru uses fuelwood fired kilns, with a production capacity of 30,000 t/y; these are described in the case study in Section 2. The plant is wholly privately owned and has a production cost breakdown as shown below. Besides fuelwood, the plant needs electricity to run its machinery. The consumption of fuel is at the rate of 18 kWh electricity and 1.71 m^3 of firewood per tonne of hydrate produced. The energy consumption translates to approximately 29 MJ/kg of product. (Note: in calculating energy consumption, the following calorific values have been assumed: fuel oil 42.3 MJ/kg, coal 29.3 MJ/kg, firewood 20 MJ/kg, 1 kWh = 4.87 kJ, density of wood = 841 kg/m^3).

It should be noted that the cost of producing one tonne of lime is less than 50% the cost of producing one tonne of cement (i.e. KSh 2,854 vs. KSh 6,094).

Special earths

There are no production plants of special earths. However most rural houses in Kenya are built using special earths. The procurement of these is generally from communal pits, which is not environment friendly and causes landscape deformation.

Table 24: Lime production and capacity utilization at the Koru plant

Year	Production (tonnes)	Capacity utilization (%)
1983	24,906	83[a]
1984	22,337	74
1985	5,557	18.5
1986	18,891	63
1987	19,113	64
1988	15,594	52
1989	24,553	82
1990	18,506	62
1991	12,047	40
1992	990	3.3[b]
1993	11,260	37.5
1994	14,400[c]	48

Source: Central Bureau of Statistics and own computations
Notes:
  a. Highest capacity utilization
  b. Lowest capacity utilization, implying that the plant was almost closed
  c. figures based on projections from monthly production between January and May.

In Kenya, there are predominantly two types of special earths: brown plastic earths and white plastic earths (kaolin). Traditionally these earths have been used as plasters on houses built with mud. The brown earths are, in addition, used for the manufacture of bricks and tiles. The manufacture of bricks on a small-scale is quite common in Western and Eastern Provinces.

Table 25: Lime production costs breakdown at the Koru plant, 1994 (per tonne)

Item	Percentage	Cost (KSh)
Wages	8.2	234.29
Raw materials	14.49	413.43
Woodfuel/electricity	19.77	564.13
Packing materials	17.24	492.00
Transport	8.96	255.68
Depreciation	9.52	271.70
Overheads	21.82	622.68
TOTAL	**100**	**2,859.94**

Source: Based on information from the management of Homa Lime Company

Cowdung/ash

In Kenya, there is no commercial production of organic binders. All the organic binders are produced by each individual household. For example the amount of cowdung a household can have depends on the number of cows the household has. Ash production is dependent on how often the cooking of food takes place in each household.

Masonry cement

There are two plants that produce masonry cement in Kenya. Nova Industries Ltd based at Athi River produces 5,000 t/y of masonry cement, while M&B Halai Ltd, based in Nairobi's industrial area, produces 9,000 t/y of masonry cement. There is no known small-scale producer of such a binder. However, it is possible that many house owners and builders are producing some kind of masonry cement, particularly a cement/lime mixture.

M&B Halai is a wholly privately owned company whose investment value is KSh 10 million. The current production cost per tonne of product is KSh 6,000. The major problem this company faces is the dependency on other plants for the provision of cement as a raw material. It is the priority of the company to produce its own clinker in future. The cost of producing one tonne of this binder is slightly less than that of producing one tonne of cement.

Nova Industries Ltd, also privately owned, has investments worth KSh 5 million. The company employs a total work force of 17 people. It costs approximately KSh 6,000 to produce one tonne of product. Like the Halai plant, the company depends on other plants for the supply of cement or clinker, and has intentions of producing its own clinker.

Masonry cement is produced by inter-grinding or mixing Portland cement clinker and gypsum with a 'non-clinker' component or in some cases two or more of such components. In the case of Nova Industries, they intergrind 25% diatomite or Kaolin with 75% OPC to produce masonry cement. This intergrinding of the components consumes approximately 782 kWh per month or 1.88 kWh per tonne of product.

The Nova product was used as a plaster for the BAT Imara Daima housing project along Mombasa Road in Nairobi. According to the BAT manage-

Table 26: Production and consumption of cement in Kenya, 1974-1994
Sources: UNIDO Study (1992) and own projections based on 1994 figures from the Central Bureau of Statistics

YEAR	PRODUCTION	CONSUMPTION	EXPORTS	REMARKS
1974	934.2	403.2	531	EXPORTS EXCEED DOMESTIC CONSUMPTION
1975	963.3	414.3	549	
1976	1064.1	436.1	628	STEADY RISE IN PRODUCTION / STEADY RISE IN DOM. CONSUMPTION / RISE IN EXPORTS
1977	1174.4	512.8	662	
1978	1151.8	541.9	609.9	
1979	1240	636	604	FALL IN EXPORTS
1980	1264.1	691.1	573	
1981	1321.2	652.6	668.6	FALL IN COMSUMP. / RISE IN EXPORTS
1982	1317	579.6	737.4	
1983	1274	517	757	ALL TIME HIGH EXPORTS AND LOWEST CONSUMPTION SINCE 1977
1984	1184.3	546.5	637.8	ERRATIC PROD. / RISE IN DOMESTIC CONSUMPTION EXCEPT FOR 1988 / FALL IN EXPORTS
1985	1164.7	649	515.7	
1986	1198.8	713	485.8	
1987	1243.3	890.5	353.2	
1988	1200.7	854.1	346.2	
1989	1316.4	1003.4	319.9	
1990	1440.9	1102.9	338	RISE IN PRODUCTION / ERRATIC EXPORTS
1991	1452	1136	316	
1992	1507.3	1158.4*	331.6*	
1993	1416.5	1088.9*	312.6*	DECLINE IN PRODUCTION CONSUMPTION AND EXPORTS
1994	1331.5*	1023.9*	293.8*	

ment, the Nova cement was used on the first 200 houses of the estate. The plasters were applied using a mix ratio of 1:4. However the use of this cement was discontinued as its life span was unknown, and it also lacked standard specifications and confirmatory tests. Blended cements, with pozzolana content of between 35% and 50%, give a generally weaker performance than OPC, but are still adequate for mortars, masonry and non-structural concrete. There is the need for a cheaper and lower quality cement than OPC. Because of this need, the Kenya Bureau of Standards is currently developing a standard for masonry cements.

Rice husk ash

Rice husk ash is not produced commercially in Kenya, however it is possible to produce this ash using low-cost technologies. At the current production levels of rice in the rice growing areas, it is possible to produce in excess of 1,600 tonnes of rice husk ash per annum. If the growing of rice in the Tana River delta were to be done, it is projected that over 5,000 tonnes of rice husk ash can be produced per annum. One of the other possible by-products of rice which could be used for ash, is the straw. The straw represents a considerably larger quantity (per tonne of paddy) as compared to the husk. However it should be noted that straw for the use of ash as a binder gets competition from other uses, e.g. production of straw boards, paper and animal fodder.

Gypsum

The procurement of gypsum is either as a natural material or as an industrial by-product. No records exist which show how much gypsum is produced by industries. However, there are large deposits of this binder in many parts of the country. As discussed earlier, these deposits are found at El Wak, near Malindi, near Garissa and between Athi River and Kajiado.

Gypsum production figures have been very variable, ranging from nearly zero to about 11,000 t/y in the early 1980s. But these figures exclude gypsum used in the manufacture of Portland cement which requires 4-5% of gypsum, which is interground with the clinker. Although this may seem to be a relatively small percentage, it still adds up to between 50,000 and 70,000 t/y for the two factories together. In addition to gypsum being used in the manufacture of cement, it is also used in fertilizers, as a filter in various materials such as paper and paint, and in the manufacture of 'plaster of Paris'.

Binder consumption and distribution

Cement consumption

The current consumption level of cement in Kenya is nearly 1 million t/y. All the cement consumed is produced by the two cement producing plants of Bamburi and Athi River. In addition to this domestic consumption, cement is exported to the Arabian gulf region and the islands in the Indian Ocean. There are also small quantities of exports to the PTA region, particularly Uganda, Tanzania, Zaire, Rwanda and Sudan.

BPCC currently produces approximately 1.1 million t/y of cement, of which almost 460,000 t/y are exported, giving a net local consumption of 640,000 t/y. Nearly all the 340,000 t/y of cement produced by EAPCC is consumed locally, however there are occasional exports to some PTA member states. There are no known imports of Ordinary Portland Cement (OPC) to this country. The BPCC production level seems to be fairly constant throughout the year, except when there is some marginal drop during the annual maintenance period. On the other hand the EAPCC experiences low sales in the months of February, March, April and August/September. This is due to the heavy rains in February-April and the annual maintenance of August/September.

Table 27: Cement consumption per capita per province (kg)

Province	1989	1994
Nairobi	269.5	196.4
Central	39.3	34.5
Coast	90.8	70.6
Eastern/Northeastern	10.2	16.8[a]
Nyanza	31.0	22.7
Rift Valley	43.7	25.4
Western	19.8	22.1
Mean	49.6	40.0

Source: Own calculations based on 1989 Population Census and Cement consumption figures from EAPCC
Note: a. Despite the prevailing economic conditions, this province shows an increased consumption due to the many donor financed projects.

Consumption of cement in Kenya has been recently growing faster than its total domestic production. This was happening on account of permanently shrinking exports. In the period 1986-1991, production of cement grew by approximately 21% at an average rate of about 4% per annum. At the same time consumption grew at an average rate of over 10% per annum, by almost 60%. After sudden upsurges in 1987 and 1989 the rate of growth of cement consumption has recently had a declining tendency. In 1991 consumption grew by only 3% and in 1992 it grew at approximately 2%. In the same period (1986-1991), with the exception of 1990, exports had a steady negative rate of growth, falling in the whole period by approximately 35%. Table 26 shows the production

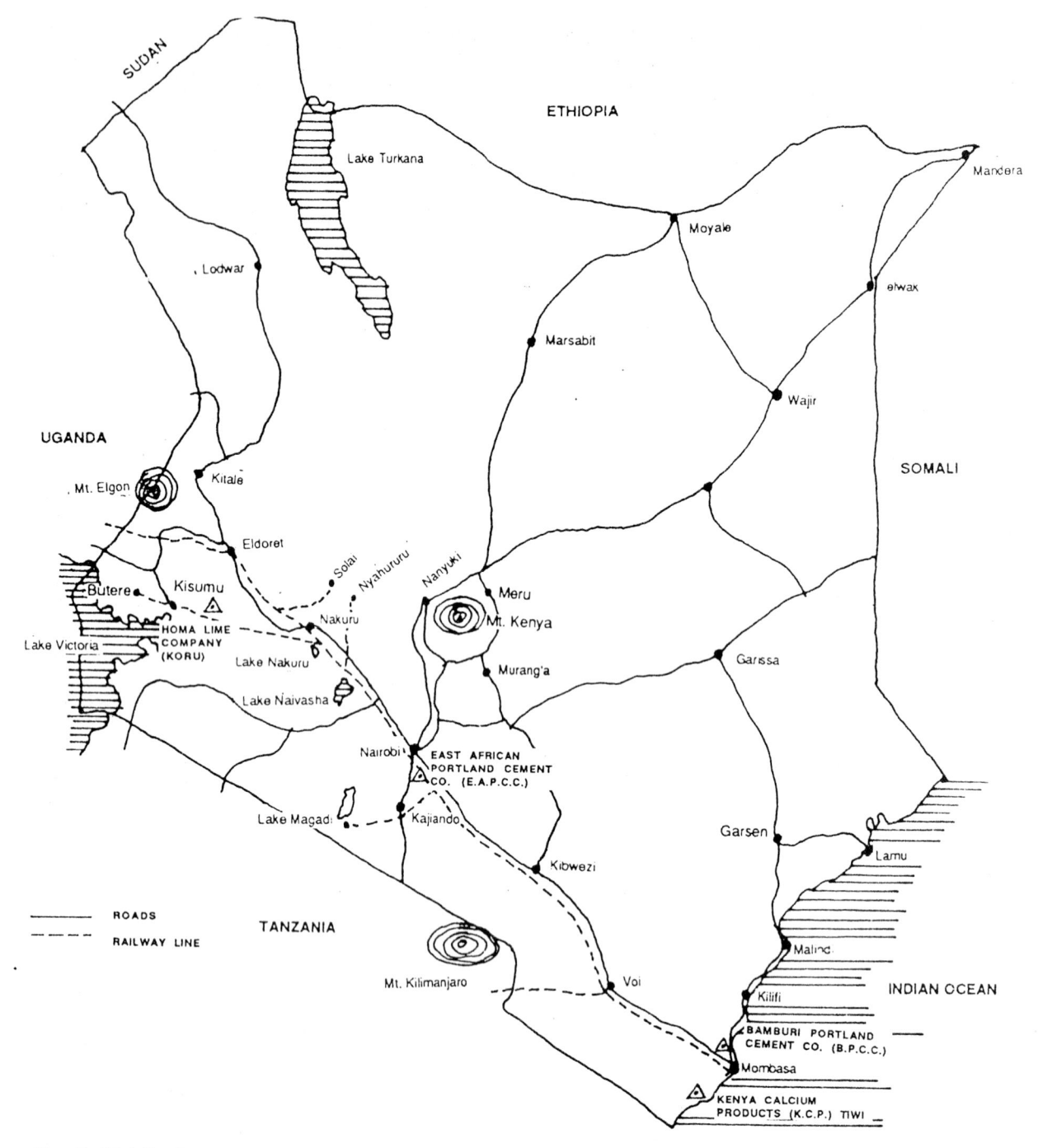

Map 5: Distribution network and location of cement and lime plants in Kenya

and consumption levels for the period 1974 to 1994.

The rate of growth of domestic cement consumption varies between 3% and 8% per annum. If the consumption grows at the lower level of 3-4%, a critical situation will occur around 1995 and a shrinking export potential at the end of the current decade. Assuming high rates of consumption growth (7-8%), there will be practically no cement left for export purposes in 1994-95. This situation will improve slightly after the new production line is installed at Athi River plant in 1996, but again Kenya becomes a net importer of cement by 1997 at the latest.

The importance of the centrally located markets is highly pronounced. The Nairobi and Central Province sector of the domestic market accounted for 47 - 49% of the total domestic sales of cement in Kenya over the last decade. The share of the Nairobi market increased from 35.2% in 1983 to 37.1% in 1991. The importance of the coastal region market continues to be rather limited, in the order of 15%. However the Eastern and Northeastern market share of the total domestic market increased from 0.07% in 1984 to 6.5% in 1991. On the other hand the Rift Valley regions market share declined from 20.8% in 1987 to 16.5% in 1991.

Table 28: Cement consumption per capita in Kenya, 1983-1994 (kg)

Year	Consumption
1983	25
1984	28
1985	32
1986	34
1987	43
1988	46
1989	49.6
1991	56
1994	40

Source: Own calculations based on data from the CBS and EAPCC

The two cement producing plants have a near monopoly in the marketing of cement. In order to minimize competition among themselves, they have agreed on some system of sharing the domestic market sales of cement. Except for the coast and Northeastern provinces, both producers are visibly present on all other markets. It is evident that BPCC supplies more than 50% market share of each region including Nairobi and Central Provinces.

The cement export market

Traditionally, almost all Kenyan cement or clinker export sales have been by BPCC, due to its favorable location on the coast. Only marginal quantities have been sporadically exported by EAPCC. Most of the export has been to Reunion, Mauritius and Uganda; relatively small quantities of cement have been delivered to numerous customers in the Indian Ocean, Arabian Gulf, and inland Africa.

Table 29: Exports of cement and clinker from BPCC, 1989-93 (tonnes)

Year	Exports
1989	317,488
1990	338,342
1991	324,157
1992	413,157
1993	622,616

Source: Central Bureau of Statistics (CBS)
Note: Approximately 50% of exports are clinker.

Lime consumption

The lime produced by the Homa Lime plant at Koru is currently consumed at the rate of approximately 10,000 t/y. Of this tonnage, 5% (500 tonnes) is for the export market, mainly to Uganda. The Kenya Calcium Products (KCP) plant at Tiwi, has a production capacity of approximately 25,000 t/y of lime. However due to the erratic market of the last three years, the actual output has fallen from 60% capacity to 40%, i.e. 10,000 t/y.

Table 30 shows that the consumption per capita is approximately 1kg, which is much lower than that of cement at 40kg.

Table 30: Lime consumption per supplier (tonnes) **and per capita** (kg) **in Kenya**

Year	Homa Lime	KCP*	TOTAL	per capita
1991	11,700	15,000	26,700	1.2
1992	6,327	12,500	18,827	0.8
1993	8,806	11,250	20,056	0.82
1994	14,400*	10,000	24,400	0.95

Sources: Information sourced from Homa Lime Company and (*) own projections based on available data from both Homa Lime and KCP.

Of all the lime produced by the Homa Lime Company, the approximate breakdown of its use per month is shown in Table 31.

The reduction in sales between 1991 and 1992 of approximately 45% was due to the poor performance of the general economy. In addition to this, there was an absence of road construction. This situation however is expected to improve in the next few years.

Table 31: Uses of lime from Homa Lime

Industrial use	Volume (tonnes)
Chemical industry (hydrate of lime)	600 (50%)
Building industry	100 (8%)
Road construction	200 (17%)
Animal feeds (stock feed)	100 (8%)
Agriculture (by-products)	200 (17%)

As discussed earlier, approximately 5% of the lime produced is for the export market. The export figures have steadily declined in the last three years. This can be attributed to the drop in production capacity at KCP and Homa Lime, and possibly a lack of marketing strategy. There is also the competition from cheaper lime and other alternatives from other countries.

Masonry cement

The current consumption of masonry cement produced by M&B Halai Ltd, is approximately 9,000 tonnes per annum. Nova Industries' market share is approximately 5,000 tonnes of cement per year, making a total annual consumption of 14,000 tonnes. All this cement is consumed by the domestic market. The market price for the Nova cement is KSh 205 per 25 kg bag, while that for the Halai cement is KSh 195 per 25 kg bag. The current market situation is such that nearly all the masonry cement is used for wall plasters. The two large constructions where this product has been used extensively are the Halai Housing Estate in Nairobi South 'C' area and the Imara Daima (BAT) Housing Scheme along Mombasa Road in Nairobi.

Other binders

Other binders, which include special earths, cowdung and gypsum, have no documented data on consumption levels. In fact there is no known case where gypsum has been used as a binder, despite the large deposits countrywide. In practice, special earths and cowdung are widely used in most rural areas.

There are five main binder types in use in Kenya today, which use various types of distribution systems to reach the consumer. The high technology binder (cement) and the intermediate technology binder (lime) which are centrally produced in factories, depend on the sea, rail and road network for their distribution. The indigenous technology binders of special earths and cowdung, which are locally produced, rely on local manual distribution. Small-scale production of lime in parts of Wajir and Lamu also depends on manual local distribution.

The newly emerging binders of PPC and masonry cements, like OPC and lime, also depend on the rail and road network for their distribution. In addition to the road and rail network, cement distribution for the export market in the Indian ocean region depends on transportation by sea. Map 5 shows the road and rail network and the location of cement and lime producing plants countrywide.

Cement distribution for the domestic market

There are four marketing companies which have the right to buy cement directly from the producers and then market the product through their agents to the final consumer. These companies are the Kenya Cement Marketing Company (KENCEM), the Kenya National Trading Corporation (KNTC), the Kenya Grain Growers Union (KGGCU) and Nzama Kuu. The internal division of the national market among the marketing companies is balanced, with each company servicing approximately 25% of the domestic market. KGGCU has not been active for some time, making the market share of the three remaining companies increase to approximately 33%.

Typically a final consumer buys cement from a marketing company's agent (if the agent acts as a wholesaler and retailer at the same time), or from a retailer who is in turn supplied by an agent. The only exception to this rule is the coastal province where the customers may (and they usually do so) pick up cement straight from the factory. The EAPCC also sells cement directly to large consumers involved in building construction. In the remaining parts of the country, a direct sale is not possible and most of the final cement depots are owned by the retailers who become an unavoidable link in the distribution system. The only marketing company that owns its own, fully self-sufficient sales network, including depots, is the KNTC.

In a few cases of extremely distant destinations (e.g. Kisumu in Western Kenya) an additional surcharge in the order of a couple of shillings on a 50kg bag is added to partially cover the extra transportation costs. The prevailing situation of cement prices is such that the more distant customers are being subsidized by those located close to the cement plants.

Lime distribution

The lime produced by Homa Lime Company at Koru and Kenya Calcium Products (KPC) at Tiwi relies on road and rail transport for the domestic distribution. The two main commercial producers of lime are also the main suppliers, however some of the cement marketing companies like KNTC also market lime. To complement the above, the two lime companies have appointed dealers in Mombasa and Kisumu. Road transport is quite expensive, however the cheaper transport by rail is extremely unreliable, as there is a constant shortage of wagons and breakdowns. This additional transportation cost is passed to the consumer, making lime more expensive than it actually should be. Another problem faced by KCP is that, being situated on the south coast, they have to rely on the use of the ferry service to reach the rail head at Mombasa. This is costly because of delays, making it difficult to make more than 2 lorry trips a day for a 30km round trip.

It should be noted that the Koru plant produces grey hydrate of lime with 55% CaO, while Kenya Calcium Products (KCP) produces white hydrate of lime with 65% CaO. The ex-works price of the KCP product is KSh 4,800 per tonne. This is because KCP product is superior to the Koru product. In practice, in order to make the Koru product more acceptable in the building industry, it is normally blended with the KCP product. This reduces its greyness and makes it much whiter.

Other than the commercial producers of lime, there are also small-scale producers of lime in Lamu and parts of Northeastern province. The distribution of lime produced in these areas is mainly manual, or by carts drawn by animals. The distances involved are short as the small-scale production is meant for local consumption. This production also relies on indigenous technologies, making the final product cheap and affordable by most low-income earners.

Distribution of other cements

The distribution of Portland Pozzolana cement is similar to that of Ordinary Portland cement. Masonry cement has yet to be popularized on the market, the current practice is therefore for the consumers to buy directly from the manufacturers. This eliminates the additional handling costs normally charged by middlemen in a large network.

Future market trends
Cement
It is estimated by UNIDO that demand for cement in Kenya be approximately 2.8 million t/y by the year 2000. This reflects a short fall of 1.25 million t/y when compared to current production levels. The EAPCC is currently undertaking an expansion project, which is expected to be completed by 1996. This will increase the plant's production capacity to 540,000 tonnes per year. This production added to Bamburi's 1.2 million tonnes will give a total country-wide capacity of 1.74 million tonnes.

Based on the above scenario, it is imperative that alternative sources of cement be put in place before the year 2000, otherwise Kenya will become a net importer of cement. It should however be noted that the proposed third Kenya Cement project in Kerio Valley, which was to be commissioned by 1996, has yet to take off the ground. Severe cement shortages are already in sight and could jeopardize further infrastructural, industrial and housing developments in the country, if not counteracted in time.

Although the export of cement from Kenya is a valuable contribution to improvement of the balance of trade, it is in the long run only desirable if it is also profitable. It is therefore important to develop the cement industry to a high level of efficiency for enabling the production and sale of cement at prices which can compete with prices offered by exports from other countries.

The two cement factories currently only produce Portland Pozzolana cement; occasionally they produce Ordinary Portland cement. The Kenya Bureau of Standards in May 1994 launched a new standard for PPC: KS 02-1263. This standard specifies requirements for composition, manufacture, strength, physical and chemical properties of PPC. This standard allows for addition of pozzolana from 15-30%. It would have been expected that consumer prices would decline with the production of PPC. However consumer prices have remained stable, the argument put forward by the cement companies being that since the pozzolanas are mined and processed, their introduction has little effect on the price structure.

The main argument from Bamburi for not reducing the cement price, although more pozzolanas are added, is that the pozzolanas used (tuffs from Nairobi) require more grinding than the cement clinker. Whereas energy is saved, because pozzolanas do not require firing, additional energy is required for the grinding, which, according to the factory, evens out the balance. However, it could be that the pozzolanas used are of rather poor quality, and therefore need so much grinding, that they do not really offer an economic option. The Bamburi plant may have to identify other sources of pozzolanic materials, preferably in dry or arid regions. Such materials are likely to be glassy, and therefore requiring less grinding and offering a more economic option. This can then be translated into a reduction of price on the final product. It should also be noted that by introducing 15-30% of pozzolana to the clinker, the production capacity of cement is increased by the same percentage without any additional investment in the plants. Thus the maximum production capacity of BPCC would be increased to 1.56 million t/y annum and that of EAPCC to 455,000 t/y. This will result in a total annual production capacity for the country of 2,025,000 tonnes.

Lime
Currently, the production capacity of lime is in excess of the demand. However it is anticipated that with the increasing cost and demand for cement, many home builders will turn to alternative binders. The Homa Lime Company which currently operates at only 30% of its capacity will require to operate at full capacity. The KCP at Tiwi, which currently operates at 40% capacity, and produces the superior white lime, will also have to operate at full capacity in order to cope with the increased demand for lime. In addition to operating at full capacity, it will be necessary to venture into the production of lime pozzolana, as a cheaper alternative to lime.

As mentioned earlier, a possible location for the production of lime pozzolana is Londiani. It should also be possible to produce lime pozzolana in the Voi/Sultan Hamud area, where there are abundant deposits of limestone. Pozzolanas for production in this area can be readily mined at Athi River. In any case since no thorough investigation has been carried out, there is a likelihood of identifying high quality pozzolana in this region which is dry and semi-arid. Namanga and El-wak are other possible production locations.

Special earths
The use of special earths in future depends on the fact that currently it costs almost nothing. However the procurement of earths from communal pits causes landscape deformation. The cost of cement and stone is constantly increasing, this will encourage more people to turn to using burnt bricks, which are manufactured from brown earths. It will be important to develop a comprehensive program which will encourage the use of earths, and at the same time rehabilitate the resultant pits, from which the earths are mined. With increasing acceptability and further research on the use of special earths, there is bound to be a rise in the demand for this binder.

Cowdung/ash

The future market trends on the use of organic based binders will depend entirely on population dynamics. The pressure on land for the settlement of people will minimize the land available for animal husbandry, and ash production. This will necessitate the use of industrially produced binders. However if the population growth stabilizes, there will be increased production of organic based binders, making them a cheaper alternative.

Pozzolanas, masonry cement, rice husk ash and gypsum

Large deposits of natural pozzolanas are to be found in many parts of the country as mentioned earlier. These deposits can be mined and blended with lime or cement. In addition to natural pozzolanas, burnt clay, bagasse ash, and coffee husk ash can also be sources of artificial pozzolana. Small-scale production of pozzolanic binders can be based on intermediate technologies in the various parts of the country where these materials are found.

The use of new binder alternatives is a result of the high cost of ordinary Portland cement. The two companies which produce masonry cement are desirous of producing their own clinker. This will free them from depending on the present cement plants. Nova Industries projects that they will require to produce their own clinker in the future. This will not only create employment opportunities, but will also reduce the reliance on ordinary Portland cement.

The production of rice husk ash depends on the production of rice. There has been no significant increase in the production of rice in recent years. However there are proposals to start rice production in the Tana River delta. Experiments in this delta show that rice plantations are devoured by quiver birds. This problem will have to be solved before any meaningful rice production can be achieved.

In future gypsum may find a wider use in the building industry, as it is readily available as a natural material or as an industrial by-product. It is also cheaper than lime or cement, since it is produced using less energy and equipment. It will also continue to be used in the manufacture of plaster of Paris, in fertilizers, and as a filler in paper and paint.

Development issues

Cement

Quarrying has a major impact on the ecological balance of a given region, since the flora and fauna of the quarried site are forcibly removed. Quarrying leaves behind unsightly holes, causing landscape deformation and in some cases deforestation. The EAPCC uses furnace oil, electricity, petrol and diesel as the main types of energy. The furnace oil is used for clinker production, and is consumed at the rate of 120 tonnes per day. Electricity is used for running the static plant i.e. crusher and other fixed machinery while diesel and petrol are used for the mobile plant, basically trucks and vehicles. On the other hand the Bamburi plant consumers 115,000 tonnes of coal and 18,000 tonnes of fuel oil per year.

The production process of cement is not environment friendly as it is essentially dusty, this necessitates strict enforcement of environmental protection and works safety regulations, e.g. installation of dust filters and control of CO_2 emission to the atmosphere. The need to control the consumption of fossil fuels and the pollution of the environment with harmful waste materials is a major concern with reference to the production of cement. The cement industry is one of the main consumers of energy and at the same time one of the biggest industrial sources of CO_2 emission and therefore called upon to reduce its energy consumption to the lowest possible level.

The production of cement requires highly skilled personnel particularly those who work in the laboratories and the factory. These will include chemists, geologists and process engineers who are required to ensure high levels of quality control. Skilled and semi-skilled persons will be required to man the operations in the quarries. To enhance the competence of the personnel, there is a need for training and retraining personnel on matters pertaining to the cement industry, so as to keep abreast with current global production practices. The cement industry generates a substantial number of jobs through direct employment at the production plants, distributorship and the entire construction industry.

Lime

Lime is produced on a large scale based on intermediate technology by two plants, Homa-Lime Company plant at Koru and Kenya Calcium Products at Tiwi. There is also small scale lime production based on indigenous technology in Lamu and Northeastern province. The quarrying of limestone causes landscape deformation and soil erosion due to rain and wind action. The continued use of wood fuel at this plant also enhances deforestation. At KCP for instance, 90 cubic feet (approx. 2.5 m^3) of wood fuel is used to produce 1.3 tonnes of lime.

The production of Lime in Kenya requires the following annual licences:

- Manufacturers occupation licence, used for export transactions
- Manufacturers licence
- County Council occupational licence

Regulations on wages follow the guidelines imposed by the Ministry of Labour, also in force are a collective agreement with the trade union. Protective clothing where appropriate is supplied, and employees are also regularly examined by medical officers form the Ministry of Health (factory inspectorate section). Pollution of the atmosphere by the emission of smoke is a major problem in the production process. However it is in the interest of the company to minimize this emission for reasons of fuel efficiency although the nature of fuelwood used means that the content of toxic wastes is low.

Re-use of quarries is not mandatory but is possible in the case of Homa Lime, because of the quick regeneration of vegetation. The quarries may then be used for grazing cattle or planting trees.

The Homa Lime plant at Koru is the least efficient, requiring 17.6 man-hours to produce one tonne of product. On the other hand BPCC plant is the most efficient requiring only 1.76 man-hours to produce one tonne of product. In general it is evident that the cement producing plants are capital intensive, while the lime producing plants are labour intensive. The capital intensive mode of production normally requires highly skilled personnel, whereas the labour intensive mode is adequately provided for by unskilled and semi-skilled personnel.

Cowdung/special earths

The procurement of cowdung does not seem to have an immediate effect on the ecological balance. However the continued use of cowdung, would eventually deny the soil essential manure which enhances soil regeneration. Cowdung is yet to be commercialized, therefore its production and distribution process can not be easily quantified. If demand for cowdung were to increase, commercialization would inevitably creep in; making the procurement of cowdung begin to have an environmental impact.

The quarrying for special earths from communal pits has a similar ecological impact like the quarrying of limestone. Special earths do not need to be processed before use, and their distribution network is limited. The procurement of special earths therefore has little impact in terms of energy consumption and other related variables. The use of special earths generates some employment, however the level of skills that go with its application are minimal. Traditional knowledge passed on from one generation to the next will suffice when working with special earths.

Portland pozzolana cement and masonry cement

The processing and distribution of PPC and masonry cements would generally have similar characteristics to OPC. Masonry cements would normally require less energy in their production than OPC. There are however some pertinent issues worth noting about these cements. Masonry cement is made by mixing of Portland cement clinker and gypsum with a 'non-clinker' component, or in some cases two or more of such components. 'Non-clinker' components with pozzolanic properties contribute to the strength development of the mortar or concrete, and masonry cement based upon the use of a pozzolanic component will therefore often produce high final strength, although the initial strengths are low in comparison with strength obtained with ordinary Portland cement. For some applications such as the construction of massive concrete structures, blended cements are often preferred over OPC because of lower heat of hydration during the curing period. Examples of this are the Turkwell Dam and the Kiambere Dam. The production of blended cements increases the plant capacity in terms of tonnes of cement, reduces cost of production especially the consumption of fuel oil for clinker production, and also extends the useful life of the limestone reserves.

Conclusions and recommendations

- There has been a general decline in exports since 1983, this could be attributed to the increased domestic consumption, without a corresponding increase in the production capacity.
- The local distribution of binders, using human labour and animal drawn carts, reduces the number of middlemen in the distribution network: This cuts off additional costs accruing from overheads and handling charges, that middlemen charge.
- Blended cements are much cheaper than OPC, however they have not been popularized, and are therefore out of reach to most consumers. Only large contracting firms have access to these cements. Producers of blended cements should therefore make an extra effort in marketing their products.
- The reduced costs in the production of Portland Pozzolana cement have not been passed on to the consumer. Ways and means should be devised where PPC can be produced at local level, so that the reduced cost of production benefits the consumer directly.
- Investigations should be done to ascertain the possibility of setting up a new plant in the main market area of Nairobi and close to pozzolana deposits, with possibilities of transpoorting clinker from Mombasa to the new plant for the manufacture of pozzolanic cements.
- Although considerable reserves of limestone exist in Kenya, the high quality deposits are rather scarce, which means that a cement plant like Athi River must base its production on the quarrying of limestone from several deposits, each with limited reserves. Future possibility of separating clinker and cement manufacture should be looked into.

Nearly all the raw materials required for binder production are either mineral based or organic based. There are, however other raw materials which are by-products of either organic or non-organic materials. The production of binders in Kenya can be classified into three main categories: the high technology, intermediate technology and indigenous technology modes of production.

The production of high-technology based binders is part of the colonial legacy of this country. Raw western technology was imposed on Kenya, without any meaningful circumstantial modifications to suit the local conditions. This mode of production thrived when there was an abundance of raw materials and cheap labour.

The intermediate technology mode of production is also part of the colonial legacy of this country, and can be attributed strongly to the white settler community. In this mode of production, western technology had to undergo through a transformation, so that it could be compatible with local conditions. This mode of production is basically a hybrid between high technology and indigenous technology modes of production. The indigenous technology mode of production is based on historical continuity. This is a technology which has been developed over a period of time and passed from one generation to another. It thrives because it is people based and has almost no monetary cost to it

The rapid population growth in this country, decreases the raw materials available for binder production. It also puts to test the modes of binder production. The high-technology mode of production becomes very expensive and out of reach to a high proportion of the population. On the other hand, because of changing value systems, cultural acceptability and density of settlements, indigenous technologies become marginalized.

The question to be answered, in these changed circumstances, is what is the appropriate technology necessary for the production of binders today? The possible answer is that production technologies will increasingly be intermediate technology based.

The most feasible scenario in resolving the issue of scarcity of raw materials would probably be as follows:

• Make a comprehensive study and map out all the raw materials that can be used in binder production. At the end of this exercise, a national land use master plan would have to be drawn up.

• After compiling the information on the raw materials, the ways and means of exploiting them should form part of the regular 5-year development plans. In addition, an enabling environment has to be created that will facilitate the generation of appropriate technologies compatible with the production of appropriate binders. These appropriate technologies will have to be popularized.

• Finally, the culture of application of the resultant binders will have to be disseminated. Here again NGOs and CBOs will have to play a central role. In this context, the importance of the construction of demonstration units cannot be over emphasized.

These are various types of production technologies that can be used in the production of binders. The high technologies are usually capital intensive and normally require highly skilled personnel. On the other hand, indigenous technologies tend to be labour intensive and require semi-skilled and unskilled persons. High technology based means of production by their very nature are expensive. Their application results in expensive products which are unaffordable to most low-income earners.

Affordability and social acceptability of products will determine what kind of products can be produced, and therefore the technologies that go with the production of these products. Products like special earths are currently seen as a poor man's choice. However if they are transformed, may be by blending them with other cementitious materials, they might begin to be more socially acceptable.

Blended special earths should be able to provide better environmental protection, particularly protection from wind, rain and solar action. The area of energy consumption is also very important, in so far as binder production is concerned. Production of binders where the level of energy consumption is low will naturally produce less expensive products. These production technologies are normally intermediate in nature. Intermediate technologies are going to have a wider application in the future, since the demand for binders is bound to increase as the population increases. Because of urbanization, consumption levels of binders is bound to rise at a higher rate than population growth, making the need for higher production capacities even more critical. For example, the population growth rate of Nairobi is 10% per annum whereas the national population growth is 3.4% per annum.

References

African Centre for Technology Studies, "Development of the cement industry in Kenya", 1991.

Lavalin International: "Feasibility study on the third Kenya cement plant located at Sebit", Final volume 1, Feb. 1990.

Mills, P.A, *Materials of construction*, 6th edition, John Wiley and Sons. Inc. New York, 1955.

S.B. Tietz and Partners: "Proposal for the location of reactive pozzolana- Koru, Kenya", 1985.

Schreckenbach and Abankwa: *Construction Technology for a Tropical Developing Country*, GTZ for the department of Architecture University of Science and Technology Kumasi-Ghana, 1981.

Smith, R.G.: "Rice husk ash cement: Small-scale production for low-cost housing", in: Proceedings of the international conference on low-cost housing for developing countries at Central Building Research Institute, Roorkee, India, 1984.

Smith, R.G: "Alternatives to OPC", Overseas

Building Note (OBN) 198, Building Research Establishment, Watford, UK, 1993.

Stulz,R and Mukerji, K.: *Appropriate building materials,* SKAT and IT Publications, 1988.

Swamy,R.M.(ed): *Cement replacement materials,* Blackie and Sons Ltd(Surrey University Press), 1986.

Tuts,R: "Pre-feasibility study on the use of Rice Husk ash as a cementitious binder in Kenya", 1990.

United Nations Industrial Development Organization (UNIDO): "A study on the present situation and proposal for development strategy of the cement industry in Kenya", 1992.

Zimbabwe

Peter Tawodzera

Zimbabwe's present day economy is still trying to recover from a recession intensified by 1991/92 drought, in the framework of an Economic Structural Adjustment Programme (ESAP) and trade liberalization. Whether the structural adjustment programme has done anything to strengthen the economy is still subject to debate.

The economy, which is estimated to have achieved a steady growth rate of between 3.5-4% over the current year, still faces a number of challenges which include:

- Inflation which stands at about 20%. Average real wages have tumbled by almost a third since 1990 and now stand at their lowest point in a quarter of a century;
- The average price index of building materials, including cement and allied products, using the 1980 prices as base figures stands at 1021.2;
- Unemployment is estimated to be more than one-third of the total workforce, unacceptably high by any standards;
- The national housing backlog has doubled over the last decade and now stands at 700,000 units. The country's performance in trying to clear this backlog is far from satisfactory. It is estimated that there are about 1.2 million homeless people in Zimbabwe;
- Rising energy costs. Electricity tariff charges have increased by almost 18% in less than 18 months. Energy consumption figures per annum in Zimbabwe stand as follows: 1.43 million tonnes of coal (this is 100% locally available) 8.97 million tonnes of wood fuel and charcoal (this is a 100% local resource) 8,309 GWh of electricity (23% of this is imported). The deliberate effort to reduce consumption of fuelwood to avert deforestation will no doubt place more demands on coal and electricity supplies;
- The construction industry is failing to recover fully from the recession due to high interest rates, the high cost of building materials, and a shortage of loanable funds particularly with regards to individual developers. The industry has however witnessed a steady surge which is expected to get to its peak by 1996. Zimbabwe's construction output as a percentage of its GNP is between 2 and 4% and when compared to other developed countries (e.g. Japan 18%, West Germany 8% and France 10%) indications are that there is a dire need for increased construction output.

Cement production

Cement is undoubtedly the most popular binding material used within the country. Whether it is low cost remains highly questionable. Cement production in Zimbabwe is monopolized by two companies, Circle Cement and United Portland Cement (UNICEM), which operate three plants. The major plants are about 580 km apart and the third plant is about 150 km away from the biggest of them all. Combined cement production per annum is currently about 1.2 million tonnes. Thus the per capita production is about 115 kgs (2.3 bags of cement).

The production of cement in Zimbabwe is biased towards large-scale, capital-intensive plants. However, in recent years with rising costs of energy and distribution, it has been difficult to realize the benefits of economies of scale, nor to bridge the demand-supply gap, in a country where tariff equalization measures are non existent. Moreover a significant amount of the present day equipment is obsolete; some machines dating back seven decades and some of the technology like the semi-dry process is not energy efficient. This causes frequent breakdowns and high maintenance costs and in turn influences the price of the final product. The fact that most of the equipment is imported means that lengthy delays could be experienced in repairing breakdowns if there is no local capacity to rectify the problem, as was the case at one of the plants in 1989.

The Zimbabwe cement industry has been operating at an average of about 90% of installed plant capacity with 11% of production output being exported. This is a highly risky operating margin which would be difficult to contain in a major breakdown case, even without export demands.

Dust pollution at one of the cement plants which was recently equipped with electrostatic precipitators is now 111 milligrams per cubic metre of air, compared to US standards of 30 milligrams per cubic metre of air, a standard which would improve visibility and prevent health effects. Prior to the installation of the electrostatic precipitators dust pollution was in the region of 700 milligrams per cubic metre of air.

Plans are underway to install a 600 t/d cement plant at the cost of about $50 million, which would eventually employ about 225 people. The equipment earmarked for this plant will be almost 100% imported. This investment works out to about $0.22 million per work place. The materialization of this project which is scheduled to be around early 1997, is unlikely to change the present scenario in terms of demand and the price of cement.

Annual demand for cement is anticipated to grow at a rate of 12% per year. The current retail price of cement in Zimbabwe is about $4-$6 per 50kg bag.

Lime production and use in Zimbabwe

Despite its widespread application, cement is not ideal for all applications. For example for low rise low-cost buildings it is much stronger than required: a lime based cement would probably be lower in cost and would perform satisfactorily.

Zimbabwe is endowed with numerous small to medium limestone reserves, limestone occurring in almost all the major geological units. There are presently well over 248 known limestone deposits in Zimbabwe. The deposits of the archean green stone belts, generally found in lenticular masses in close association with banded limestones, have been the main supply source which has met local demands. The problem connected with the Zimbabwean limestone deposits is not quantity, but maldistribution and perhaps quality; the location of suitable limestone does not coincide with high population densities. Remoteness and distance from the railway are serious obstacles to exploiting many promising deposits. However this problem could easily be converted into an opportunity if small scale exploitation of these deposits were considered for local consumption.

The problem of quality is, arguably, a relative one. Although appreciable reserves suitable for cement manufacture, building lime and agricultural purposes exist in the country, chemical grade limestone, which is acceptable to the ferrochrome industry, the sugar industry etc, is scarce.

In Zimbabwe there are ten principal uses of limestone and the lime obtained by its calcination: first, as a flux in the iron and steel and ferrochrome industries; second, as an aggregate or soil stabilization agent in road construction; third, as a cement extender in construction; fourth, as a soil conditioner/pH modifier in agriculture; fifth, in animal feed as a mineral supplement; sixth, as a pH modifier in the sugar industry and in water purification; seventh, as the main raw material in the production of cement; eighth, in the production of glass and ceramics; ninth, as pH modifier in gold production using the cyanidation process; and lastly, as a filler in paint manufacturing.

The main markets for hydrated lime are the sugar production industry, iron and steel and ferrochrome production, gold production and water purification. The market for lime in Zimbabwe is supplied by a medium-scale local producer, G & W Industrial Minerals, with the rest of it being imported. Cement producing plants have their own quarrying sites and tend to also supply agricultural lime (crushed ground limestone). The iron and steel plant has its own supplies of lime.

Besides the cement producing plants, there are a few other small to medium suppliers of agricultural lime, including G & W Industrial Minerals and Alaska Dolomite. Alaska Dolomite produces 30 000 tonnes per annum of agricultural lime; 40 000 cubic metres per annum of aggregate for road construction; and employs 200 workers.

The market for building lime is not regular. G & W Industrial Minerals (Pvt) Ltd meets the occasional demand for building lime by supplying industrial grade lime which is much purer than required and therefore more costly. The current price of hydrated lime from G & W Industrial Minerals is about $4 per 50 kg bag.

Limestone and lime consumption patterns in Zimbabwe are shown in Tables 32 and 33.

Table 32: Limestone consumption in Zimbabwe

Industry	Consumption ('000 t/y)
Cement	1,600
Agricultural and animal feed	100
Building and construction	4
TOTAL	**1,704**

Table 33: Lime consumption in Zimbabwe

Industry	Consumption ('000 t/y)
Iron and steel	65
Low and high carbon ferrochrome	65
Municipalities (water treatment)	12
Sugar industries	12
Mining	10
Pulp and Paper	8
Building and construction	2
Glass industry	5
Leather industry	1
Other	16
TOTAL	**196**

Approximately 100,000 tonnes of lime is imported annually for the industry, and this has to meet stringent quality requirements.

In recent years, with the escalating cost of Portland cement, there has been increased interest in the local production of lime for building. In fact, the consumption of burnt limestone in the construction industry alone is expected to rise by more than 10% per annum. To meet this anticipated rising demand, Intermediate Technology Development Group, Zimbabwe has commissioned a study of limestone deposits in Zimbabwe to assess their potential for lime production. One of the major findings according to the adopted criteria was that many of the suitable deposits were scattered and quite small in extent, making them ideal for small-scale exploitation for localized markets.

In a related development, the ongoing research work within the SADC Region (including Zimbabwe) by the Transport Research Laboratory (TRL) on low cost road construction might recommend the use of lime in low-cost road construction, which could eventually add to the increased demand for lime production within the sub region.

Because small-scale exploitation of lime has a low capital investment per unit production and the potential to create employment, as well as producing a low cost binder, plans are underway in Zimbabwe to establish a pilot project to produce lime on a small-scale basis. By comparison the cost of establishing 20 t/d cement plant is approximately $1 million, whilst the cost of investment for 20 t/d lime production plant would be less than $60,000.

One potential difficulty could be the supply of fuel (coal) to fire the kilns, the use of wood and charcoal being an unfavourable option. Thus it would be of prime importance that any kiln developed to burn the limestone would have to be fuel-efficient.

Pozzolanas

Given the abundance and diversity of pozzolanic materials in Zimbabwe, there is a great scope for developing lime-pozzolana cements with improved binding properties. The following are pozzolanic materials available in Zimbabwe:

- blast furnace slag
- pulverized fuel ash (PFA)
- burnt bricks.

Zimbabwe Iron and Steel Company (ZISCO) produces about 400,000 tonnes of slag annually. PFA is a by-product from the burning of coal in thermal power stations. Two of the four thermal power stations in Zimbabwe produce approximately 410,000 tonnes of PFA annually. Consumption of these industrial wastes is currently very small within the country. Finally, brick production is literally common place in Zimbabwe.

The costs of investment in the crushing and grinding of the pozzolanic material might prove inhibitive for a small-scale operation. Developing small-scale lime burning on the basis of a multi-product plant and introducing selective marketing might prove viable whilst providing a low cost binding material.

References:

Argossy Press: *Construction and Engineering Zimbabwe* Vol-5, Oct. 1993, pp 7-13

Internal Papers from G & W Industries P.Nhachi: "Manufacture of lime in Zimbabwe, 1985". 5 pages; The lime industry in Zimbabwe, 1992. 8 pages; Lime as a building material in Zimbabwe, 1993. 8 pages.

ITDG Zimbabwe: "Initial stage of lime production feasibility study in Zimbabwe", 1993, 113 pages.

SKAT: *Basin News*, Vol 8., 1994, pp 13-18.

Standard Chartered: *Quarterly Africa Review*, September, 1994. pp 20-22.

Standard Chartered: *Business Trends*, August, 1994. pp 1-9.

Central Statistical Office: *Quarterly Digest of Statistics*, June 1994.

Central Statistical Office: *The Census of Production*, 1989.

Dr H.Uzoegbo: "Industrial waste in building materials manufacture in Zimbabwe", 1994, 9 pages.

G. E. Radloff: "Cement Production, Supply and Demand, 1984-1993", July 1989.

United Nations (Habitat): *Small-scale production of Portland cement*, 1993, 81 pages.

B. Barber: "Calcium Carbonate in Zimbabwe", 1990, 183 pages.

SECTION 2

Lime production: Case studies

INTRODUCTION

Lime is produced from limestone rock deposits. Such deposits are quite widely dispersed in East Africa. Most commonly the limestone deposits were laid down in geological time from many billions of calcareous sea creatures such as mollusca and corals. Less common are marbles (sedimentary limestones which are metamorphosed by heat and/or pressure) and travertine (precipitated calcium carbonate). A unique limestone deposit occurs in the Tororo area of Uganda. This consists of plugs of volcanic origin. Lime as well as Portland cement is produced in the Tororo area from this deposit.

Limestones vary quite considerably in characteristics, although almost all can be used for producing some form of lime. Limestones may be crystalline or amorphous. They may consist almost wholly of calcium carbonate ($CaCO_3$), the main constituent of most limestones, or the calcium carbonate may be mixed in with other constituents. The main non-calcitic constituents are magnesium carbonate ($MgCO_3$) and silica (SiO_2). If the limestone contains between 5 and 25% $MgCO_3$ it is known as dolomite. Satisfactory lime can be produced from dolomite provided it is noted that $MgCO_3$ dissociates at some 50°C lower than $CaCO_3$, so the $MgCO_3$ component will often be overburnt. This implies a need to allow the lime to hydrate longer with water to ensure that the MgO component as well as the CaO is well hydrated. Some industrial users of lime, such as the sugar industry, are unlikely to make use of lime with more than a small magnesium component, but it can be used for agricultural purposes as well as in building.

Another relatively common type of limestone is one containing some clay, that is, a proportion of silica. On burning the silica reacts chemically with the calcium to give an additional cementitious component. Lime containing silica is called hydraulic lime because, unlike a calcitic lime, it can set under water. Hydraulic lime is stronger than calcitic lime and sets and hardens less slowly. The obvious application for such a lime is for building, and for roads. However most industrial users of lime find a silica component in lime undesirable for various reasons.

Limestones can also contain other components such as Fe_2O_3, Al_2O_3, NaO, P_2O_5, etc. Significant concentrations of such components can be undesirable in certain applications, especially for the use of lime in foodstuffs and pharmaceuticals, although they can be tolerated in building lime.

The chemistry and processes of lime production

The first stage in lime production is the dissociation of limestone into oxide components and carbon dioxide.

$$CaCO_3 \xrightarrow{heat} CaO + CO_2 \uparrow$$

Considerable heat, in fact 2.82 MJ per kg, is needed to enable the reaction to proceed.

Limestone is usually burned with fuel in a kiln and, because no kiln is 100% efficient, the heat needed to produce lime can vary from 1.2 times the above theoretical value in the most modern continuously operated optimized kiln with heat recuperation facilities to six or seven times this figure for an inefficient batch-type field kiln.

The CaO produced on burning limestone is known as quicklime and is rather unstable. It has a very high affinity to water and will even react with moisture in the air, so in humid conditions will spoil in about a day or so. When water is added directly to quicklime there is usually a violent reaction within a few minutes, with the water hissing spitting and boiling. This reaction is known as hydration and is represented chemically by the following equation:

$$CaO + H_2O \longrightarrow Ca(OH)_2 + heat$$

Because of the instability of quicklime it is usually converted to hydrated lime as soon as possible after removing from the kiln. This is done in the open air or in a purpose-built hydration unit. Hydration can also be carried out in pits, with the lime left in the pit to hydrate thoroughly and form lime putty. Lime putty is the form of lime most preferred by builders. It is being used for conservation projects on Stone Town in Zanzibar, for example. The final stages in the production process are sieving, to remove lumps of unhydrated material, and, optionally, depending on the end use, grinding and separation into size fractions using a classifier or air cyclone.

The use of lime

Lime is a very important industrial mineral and is used in many industrial applications such as papermaking, foodstuffs, pharmacuticals, paint, the iron and steel industry and fertilizers. It is also used in leather tanning, the sugar industry, and in agriculture to condition acidic soils. Some industrial applications require a very pure and fine lime, one which is likely to be produced in a medium or large-scale technologically sophisticated plant.

Possibly the most important use of lime is as a building material. In fact a high proportion of lime produced in East Africa is used in this way.

Hydrated lime acts as a binder by reacting slowly over time with carbon dioxide in the air to revert back to the hard and solid limestone material. In this respect it differs from most other inorganic binders which set and harden by reaction with water.

Lime production in East Africa

Lime is being produced, to a greater or lesser extent, in all East African countries. In some, such as Tanzania, the production and use of lime is an important economic activity in certain areas whereas in other countries, such as Kenya, production is very localized and for a specialized market, and cannot be said to contribute significantly to the national economy.

In East Africa, lime production can be considered to exist at four technical levels:

(i) Relatively highly mechanized, medium-scale, using continuously operated kilns and with a largely industrial market. Examples of such plants are Homa Lime in Kenya and Tanga Lime in Tanzania.

(ii) Low to medium mechanization, relatively small-scale, using continuously operated kilns and with a market for the lime largely in the building sector. Examples include Chenkumbi Limeworks in Malawi, Equator Lime in Uganda and the new Nyalakoti limeworks in Tororo, Uganda, as well as the SIDO kilns in Tanzania.

(iii) Low to medium mechanization, small to medium-scale production, permanent batch kilns are used and the market is mainly in the building sector. Examples include Kigeze Twimukye at Kisoro in Uganda and Hima Lime at Kilembe in Uganda.

(iv) Manual operations, low level of production, batch-type field kiln with little or no permanent structure and a market mainly in the building sector. This type of production probably exists in all the East African countries, and further reference is made to units in Zanzibar and at Tanga in Tanzania. It is difficult to quantify how much lime is produced in this way because production tends to be set up and stopped according to demand and the production site is not permanent. Most lime producers in this sector do not earn a living solely from lime production and are also involved in other activities such as farming and fishing.

Given the current extent of cement shortages in East Africa and that these shortages are projected to become more acute, there is an identified opportunity for lime to play an increasing role as a binder for building. However, existing and potential lime producers in the area do face a number of possible constraints which might mitigate against the development of the industry. These constraints are technical, social and economic and, more specifically, can be considered to be lack of comprehensive information on exact raw material reserves and quality, inefficient stone processing, high fuel costs and exclusive reliance on wood for fuel by many of the small producers, inefficient and wasteful production technologies, poor quality control, inadequate regard for health and safety practices, poor access to transport, inefficient marketing strategies, lack of access to investment capital and improved production technologies, poor site security, little co-ordination between producers and little support offered to producers from research and development organizations, inflexibility towards environmental legislation and concerns, and a general inertia within the industry towards changing circumstances.

It would be true to say that none of the producers surveyed were affected to a large extent by all the above constraints, but all would be affected by more than one of them. That is not to say that there are no producers working very efficiently and with a lot of acumen under very difficult and changing circumstances. It should, however, be noted that the constraints would be likely to more severely affect the small-scale traditional producer and it is their futures which are more uncertain than those of the medium-scale more mechanized producers.

A number of lime producers in East Africa were surveyed in 1994 and the following set of case studies were produced. Producers were, particularly, asked to consider the main problems and bottlenecks they were experiencing and how they considered their production plants could develop in the future. Significantly, most considered that prospects for increased production were good and considered expanding their facilities. The main difficulty is likely to lie in setting up a training, marketing, technical support and investment framework which will allow producers at all levels to achieve the level of expansion which they consider appropriate to their circumstances.

— Otto Ruskulis

Nyalakoti Farming and Lime Works, Tororo, Uganda

W. Okello

Nyalakoti Farming and Lime Works Co-operative Society is a small-scale lime-producing society registered in 1990 under the provision of section 5 of the Co-operative Society Act, 1970. It consists of small lime producing units which are run along family or communal lines and sometimes by individuals. It has a total of 40 members. It is located approximately 5 km from Tororo on the Tororo-Jinja Highway.

The main objective of the Society is to promote the production and marketing of lime produced by the members of the society. The major tasks often undertaken by the society include: marketing of the finished products, search for alternative techniques to improve on the production processes, securing of licences especially for the wood fuels obtained from the gazetted forest reserves, and mobilization of funds for the society.

Raw materials

The raw materials supply is from quarries located within the Tororo-Sukulu limestone complexes. Known reserves are estimated at 154 million tonnes with additional projected reserves.

The bulk of the limestone consumed is by the cement industry located at Tororo while only a very small portion, 8,000-10,000 tonnes per annum (classified as rejects) are used for small-scale lime production. The rocks are carbonatitic of volcanic origin. Analysis of a typical sample from the Sukulu Hills revealed a composition as follows:

Unburnt samples	%
SiO_2	1.2
Al_2O_3	1.1
Fe_2O_3	5.2
P_2O_5	3.5
MgO	3.09
SO_3	0.34
CaO	48.83
LOI	36.94
Total $CaCO_3$ content	85-90

Extraction technology

Basically two extraction techniques are used:
- drilling and blasting of the rocks by use of explosives.
- manual, using casual labour and simple tools and equipment such as hoes, picks, shovels, wheelbarrows, sledge hammers, wedges. In some cases relatively hard rocks are loosened by burning fires on them.

Transportation

The production units are located within 3 km of the quarries, all of which are easily accessible. Tractors, lorries and, occasionally, wheelbarrows are used for transport.

Production technology and processes

The lime kiln is a traditional hillside batch kiln with the following features:

Capacity: approx. 15 tonnes limestone per batch
Average dimension: 3.6 m height by 3.6 m diameter
Shape: generally a truncated cone tapering towards the top, but sometimes cylindrical.
Construction detail: the main elements consist of an inner lining of locally burnt clay bricks approximately 500 mm thick with clay mortar joints, and approx. 300 mm outer lining comprising limestone rubble or brick masonry with lime mortar. The latter provides for structural stability as well as for insulation.
Loading/offloading: Two doors are provided at the bottom for offloading. Loading is from the top.

The total cost of the kiln is estimated as equivalent to $1,500 in local currency.

Hydration: this is done on a level concrete floor/platform; a water sprinkler/can is used for sprinkling the water.
Sieving to remove coarse underburnt particles is by means of 4 mm and 1-2 mm wire fabric/mesh for coarse and fine sieving respectively.
Bagging in 2-3 ply paper bags of 25 kg capacity, manually filled.
Storage: often in paper bags on raised wooden platforms in a well secured adjacent store.
Transportation: wheelbarrows are often used for on-site transport while tractors and lorries are used for off-site transportation.

The Production cycle

Raw materials preparation: the stones brought from the quarry range in size between 150 and 200 mm, and are reduced to an average size of 70-100 mm; firewood is cut to an average size of 400-600 mm lengths and 300-500 mm diameter.
Batching, loading of the kiln, and firing: starting with firewood at the bottom, the kiln is progressively loaded up to the top in alternate layers of firewood and stones (approx. 0.5 m each layer) until it is full.

Fire is then lit from the bottom through the doors. As the burning proceeds a drop is realised in the height of the materials, which is promptly topped up in alternate layers as before, until the kiln is completely filled to the top. The fire takes 2 to 3 days to reach the top. The quicklime is allowed to cool for 2 to 3 days before off-loading commences.

Off-loading and hydration: the quicklime is off-loaded from the kiln through the doors onto wheelbarrows, and is spread on the hydration platforms in a layer about 150 mm high ready for hydration. Water is then gradually sprinkled on the quicklime until the stones are fully hydrated. The amount of water used depends more on experience than on any established ratio.

Sieving and bagging: the hydrated lime is initially coarse-sieved and then later fine-sieved before bagging. The rejects from coarse-sieving are either returned to the kiln for further firing or used as aggregates, while those from fine-sieving are used in concrete brick production.

Labour requirements: for every production unit a labour force of 15-20 people is required, mainly casual labourers.

Water
Potable water drawn from wells or nearby borehole pumps is used in the hydration of the quicklime.

Energy
The sole source of energy in use is local hardwood obtained from gazetted forest reserves some 40 km away, brought in trucks of 3-5 cubic metre capacities. An average of 1 tonne of firewood is required for 1-1.3 tonne of lime.

Quality of products
Generally the products are low quality hydraulic lime not suitable for chemical and industrial uses. It is difficult to maintain uniform quality.

Production capacity
The average output of each kiln is estimated at 15 to 20 tonnes per week equivalent to 60 tonnes per month per kiln.

Investment
For all units, funds are found locally through personal savings and, infrequently, by borrowing from the society. The units are owned by individuals as well as family groups or societies.

Waste
Raw material waste, that is undersized material rejected for burning, is low and estimated to be up to 5%. On burning, a loss of up to 40% is sometimes encountered, especially during bad weather conditions. These underburnt stones are usually recycled for further burning.

Product price and sales
A tonne of lime costs approximately USh 96,000 ($96). Virtually all the lime is produced on demand. The Ministry of Works and related road construction firms constitute the biggest consumers. With improved lime quality, the market is likely to

Lime production at Tororo

expand to include agricultural and industrial consumers as well.

Sales are greatly determined by the demand; random fluctuations are experienced when new road projects commence, as well as during the wet season. On average, sales have been at approximately 1,000 tonnes per month (efforts to obtain actual figures over a particular period was fruitless).

Policies and Regulations
Licences for raw materials and fuel
The society exploits the raw materials without prospecting licences issued by the Government as they basically depend on what is graded as 'rejects' unfit for cement production. A temporary licence is obtained from the Forestry Department for harvesting of woodfuel from the gazetted forest reserves.

Pricing, wage and safety regulations
Prices and wage bills are governed by market forces and production costs. Earlier attempts by the Lime Producers Association to streamline the rates proved unrealistic due to the variation in the production costs and management methods adopted by various units.

Whereas health and safety regulations are enforced by the Factories Inspectors under the Ministry of Labour, it is very rare that these services benefit all the workers. As such, it is common to see workers without protective clothing.

Main bottlenecks and problems
Production
The factors affecting production include:
- lack of a mill for milling the lime
- improper separation methods to improve on the fineness and the quality of the lime
- low firing temperatures as well as excessive heat loss basically due to use of inappropriate kilns
- frequent repair to the inner lining of the kilns; bricks used for the purpose are non-refractory and therefore cannot withstand high temperatures
- meagre working capital which limits the expansion of the enterprise
- application of simple hand tools/equipment (sometimes engagement of semi-skilled/unskilled labour force)

seasonal demand for the products.

Sales
The main obstacles to sales include:
- low quality products and therefore restricted market, and
- preference for alternative materials especially cement in the building industry.

Future plans and prospects
The Society's main targets are:
- to develop/adopt an energy-efficient lime kiln technology
- to venture into new energy sources other than fuelwood
- to produce a wide range of high-quality lime to cater for agricultural, industrial, building and road uses.

Currently an energy efficient lime kiln is being developed in collaboration with the Ministry of Lands, Housing and Urban Development (Department of Housing) and with assistance from UN Economic Commission for Africa, which is described in the next section. Meanwhile, a tree-planting campaign has been launched to ensure sustainable supplies of fuelwood.

Experimental lime kiln at Tororo, Uganda

W. Okello

This pilot project aims at encouraging the production and increased utilization of local building materials in response to low cost housing and infrastructure requirements; as well as to integrate women into housing and construction activities. It is a follow-up of recommendations signed and endorsed by Heads of States and Governments of OAU member countries at a high level meeting (26-29 April 1988, Addis Ababa, Ethiopia), i.e., 'to house the rural population and the urban poor by establishing commercial small-scale production of selected local building materials easily accessible to a majority of the population'. Accordingly, Uganda was identified as one of the countries to undertake the Pilot small-scale production and utilization of Lime, and in particular to develop an energy efficient lime kiln, an element considered most crucial in the production process for attaining high quality end products at relatively affordable costs.

The collection of the baseline data commenced in December 1992 and the design of the kiln was completed in October 1993. Actual construction works started in January 1994. Whereas production was expected to begin by the end of June 1994, it has not been possible partly due to factors such as: delays in disbursement of funds; difficulties in obtaining some materials, especially refractory cement and bricks; and bad weather conditions. The trial running of the kiln is now scheduled for 1995.

Essentially, when completed, the project will serve as a 'Demonstration Unit' as well as a 'Training Centre' for any interested party venturing into small-scale lime production. A formal memorandum of understanding has been drawn up between the Government and a local lime-producing society in the area which, on completion, will undertake the day-to-day running of the project while the Government will continue to play a monitoring role and ensure access to the site for training and research purposes.

Institutional arrangement

The UN Economic Commission for Africa (UNECA) is the executing Agency on behalf of the UNDP while the Ministry of Lands, Housing and Urban Development is the implementing agency (on behalf of the Government) in collaboration with the local lime Society, the Nyalakoti Farming and Lime Works Co-operative Society.

Investment

In addition to funding an international and local consultants, ECA has made a provision of $15,000 for construction of the kiln. A similar amount has been secured for a second kiln to be located at Kasese in the Western Part of the country. Meanwhile, over the last two years, the Government of Uganda has made available a total of USh 13- million ($13,000) towards implementation of the project. The contribution from the local society is mainly in the form of land and, to some extent, labour.

Technology

The kiln has been designed as an 'improved simple continuous shaft kiln with mixed feed'. This kiln is very similar to the Indian kiln used by SIDO in Tanzania. Its output is estimated at 5 tonnes of quicklime per day. Fuelwood is likely to remain the main source of energy, although attempts shall be made to use agrowaste as an alternative fuel.

The main components of the kiln comprise:
foundation: of masonry construction as the basic framework
basement: of masonry of good quality burnt bricks grouted with cement mortar while the protection slab is made of concrete from pozzolana cement mortar
air supply port: consisting of piers and vaults constructed of burnt bricks grouted with pozzolan cement and masonry constructed of conical burnt bricks grouted with refractory cement lime pozzolana mortar respectively
burning chamber: the interior lining is of refractory bricks grouted with a refractory cement, while the exterior is of good quality burnt bricks grouted with pozzolan cement mortar
structural wall: is of masonry of good quality burnt bricks grouted with pozzolan cement mortar
insulation: a mixture of natural pozzolana and rice husk ash
chimney: in metal, consisting of a cap, column, pavilion and a trap door
hoisting system: comprising a winch and gallows, and a loading gauge for loading of the stones and fuel.
working platform.

The production process is similar to that described under the clamp kilns in the previous section, except in this case the loading of the raw materials and fuel is through the trap doors in the chimney, and the process is continuous. A fairly skilled manpower is required compared to that for the traditional clamp kilns.

Waste, production cost and markets

The kiln efficiency is anticipated to be over 60% and as such minimal waste is expected in the production process; a relatively low production cost is targeted. With high quality products available for building, road works, chemical and industrial users, the demand is expected to exceed supply.

Main bottlenecks and problems

The main problems so far experienced include difficulty in obtaining suitable refractory bricks locally. So far only one local factory (Uganda Clays Ltd) has been identified as capable of producing the bricks; however, they are yet to perfect their performance in this area. Refractory cement is also in scarce supply.

In addition to the above, it is envisaged that there will be an acute shortage of fuelwood for the kiln unless tree planting is embarked on immediately, specifically for the venture.

Future plans and prospects

It is planned that a variety of products suitable for road works, building as well as for chemical and industrial uses shall be produced to suit the local demands and therefore reduce imports.

More lime dealers are expected to adopt the technology and possibly modify it to make use of alternative fuels other than woodfuel.

Kigezi Twimukye Lime, Kisoro, Uganda

W. Balu-Tabaaro

Kigezi Twimukye Lime Company Limited was established in 1976 to produce lime for road and building construction. It is a company owned by about 50 members who contribute various shares to raise capital for running the production of lime. The company whose headquarters are in Kabale town, is located about 65 miles from Kabale on the Kabale-Kisoro-Busanza road. It is managed by a Chairman but the lime plant is run by a manager at the production site.

Raw materials

Calcareous rock is found along the valley of the Kaku river at four localities. It usually occurs as a terrace in the river valley (possibly due to difference in erodibility of the different rock types). The river flows in a papyrus filled channel ten feet below the terrace and vegetation covers the valley and it is possible that the extent of the calcareous rock is greater than show by the outcrops.

The limestone reserves are estimated at over 2 million tonnes; they are of the following average assay values:

Unburnt samples	%
CaO	51.77
Fe_2O_3	2.08
MgO	2.03
CO_2	41.22
H_2O	2.24
Insoluble in acid	1.13

At the present extraction rate of 5 tonnes per month, the reserve can be mined for the next 1,000 years.

Production technology and processes

To obtain limestone for lime production, a suitable area of the deposit is delineated and overburden removed manually using shovels, pick-axes and crow bars. In very few instances are explosives used and that is when very hard stones are encountered and drilling is carried out. Production is also a problem during the rainy season. Once the stones are extracted, they are reduced in size to about 250 mm and are then hauled on wheelbarrows and karais by unskilled labourers to the kiln over a distance of 100 metres, for burning.

At the kiln site, the limestone is broken into lumps of size 100-150 mm by sledge hammers and stockpiled ready for loading. The firewood, bought from a nearby plantation, is also cut into logs 1 metre long by 0.3 metres in diameter.

The limestone is then loaded into the kiln: firewood is laid at the bottom first and a layer of limestone added on top, to form a layer of 30 cm thickness. This alternative sequence of loading is continued until the kiln (dimensions 3.0 m height and 4.4 m diameter) is full. The fire is then lit at the bottom and the gate closed. Lime production here is a batch process and the fire is left to burn for 7 days to complete the cycle. The processes leading to complete burning include:

1. Loading   2 days   8 men   (16 man days)
2. Burning   3 days   4 men   (16 man days)
3. Unloading 2 days   8 men   (16 man days)

After complete burning and cooling, the quicklime is off loaded and transported by karais and wheelbarrows to the hydration shed. Here the water to hydrate is obtained from the nearby Kaku river and 4.5 jerrycans of 20 litres each are used to hydrate 5 tonnes. The hydrated lime is then sieved through a 1 mm screen and packed in 90 kg bags. The sieving is done by two men at a go. The oversize, which in most cases is the underburnt limestone, is piled aside and is usually sold as agricultural lime. About 20 to 30% of the lime produced is in the form of this residue.

Energy sources

The firing system involves use of firewood which at present is obtained from a nearby plantation, but the company has also planted their own plantation which they hope to utilize in future.

Firewood costs Sh 40,000 ($40) per 7 tonne lorry and one round of firing requires 2.5 lorry loads i.e., Sh 100,000 ($100) per 5 tonnes of hydrated lime produced.

Investment

The lime company was started by individuals in Kigezi Twimukye Company who contributed shares. These shares were then used as the capital to run the lime operations. As production picked up, some of the profits obtained were ploughed back into the operations.

The company employs 15 to 20 unskilled labourers, with a manager in overall charge. There is a great need for more skilled workers in raw material identification, selection, raw material preparation, firing and hydration and sieving and quality control. A lot of information on lime production technologies is still lacking and so is the training of future operatives. But even if skilled workers were available, the low prices paid for lime would make wage rates low. Present monthly wage rates of $7 for a casual labourer to $20 for a supervisor would be a discouragement to employment of skilled labour.

Markets

The company produces up to 5 tonnes of lime per week. The lime is sold locally mainly in Kisoro and Kabale towns, 15 and 65 miles away respectively. If one collects the lime from the site, one pays Sh 7000 ($7) per bag and if delivered to Kabale, the cost increases to Sh 10,000 ($10) per 90 kg bag. Production figures for the past 18 years are shown in Table 34.

Table 34: Kigezi Twimukye's production of lime

Year	Quantity (tonnes)
1976	4.14
1977	1.24
1978	2.05
1979	1.02
1980	1.011
1984	5.02
1985	42.01
1986	47.05
1987	77.02
1988	52.01
1989	6.05
1990	61.7
1991	62.6
1992	72.5
1993	46.5
1994	20.5 up to June
TOTAL	502.58

The main buyers of lime in all these places are builders and contractors. Some of the lime is mixed with ordinary Portland cement as a binder while in other cases it is used as whitewash. Before the war in Rwanda, considerable quantities of lime used to be sold there also. It is hoped that potential still exist for future exports there since the hostilities there have stopped and need for rehabilitation of infrastructure will require binders, including lime.

The sales are affected by the rainy season when little production goes on and accessibility to the area is difficult. Because of poor quality and quantity of lime, the demand is captive; i.e areas like sugar refining, road making etc are hard to penetrate.

Prospects for the future depend on the company's ability to improve the kiln capacity and quality of the lime in order to expand into the above markets.

Policies and regulations

In order to extract limestone, a prospecting license costing Sh 28,000 (Sh 20,000 as deposit and Sh 8,000 annual fee) is paid. On certification of findings of proven reserves of limestone, a mining licence (Sh 30,000) is then required to enable commercial extraction. An annual loyalty fee chargeable on gross value of the extracted limestone, at 5% is also levied.

Firewood is usually bought from government-owned plantations and a fee for extraction is also paid. With the company now growing its own plantations, only a resource fee as income tax will be paid. There are no pricing regulations at present since the economy was liberalized and made competitive. Only quality and quantity dictate the market and pricing regimes. At present no wage regulations, including a minimum wage, apply, though there is a pending bill geared to establishing a minimum living wage for workers.

The nature of lime production is unhealthy and demands that health precautions are taken. At Kigezi Twimukye Lime Co., no mouth masks nor eye wash are provided and lack of these facilities has harmed the health of the workers greatly. Boots are not provided and lime, being corrosive, has 'eaten away' most of the workers' feet.

When limestone is mined, quarries are developed and left unfilled. These unfilled quarries are an environmental and health hazard. The department of Geological Survey and Mines, in conjunction with the Directorate of Environment has put legislation in place to ensure that these quarries are reclaimed and made safe.

Bottlenecks and problems

The main problem affecting lime production is lack of adequate transport (lorry) to carry both raw materials and processed lime. Both the extraction process and lime production are manual leading to low production and low quality lime. There is a need to acquire better mining tools. (Compressor, jack hammer drills, explosives, etc) and raw material preparation equipment (jaw crusher, sieves, ball mill, screens, balances, hydration machine). But acquisition of these facilities would require funds which are also not readily available.

The lack of a quality control laboratory also constrains production of good quality lime.

Plans and prospects

Given enough financial and technical facilities the company would like to improve the size and quality of the kiln in order to increase production of better quality lime. The company is in the process of carrying out a techno-economic study to evaluate how this can be achieved.

There are also intentions to start a pilot project on production of lime-pozzolan cement, based on the abundant volcanic pozzolans found in the Kisoro area.

Equator Lime, Kasese, Uganda

W. Balu-Tabaaro

The Equator Lime Co. Ltd owns and operates a lime production unit at Muhokya, 7km from Kasese Town along the Kasese-Mbarara Highway. The company started operations on a site that was formerly run by a local lime producer, Kabuga Lime Co. Now they have moved to a new site, where they have constructed a new kiln and are in the process of expanding into the production of lime pozzolanas using their lime and the Bunyaruguru volcanic ash, 32 km away along the Kasese-Mbarara highway.

Raw materials

The Muhokya deposit contains a marly limestone. The rock is compact, light grey with small cavities and shells of fresh water mollusca and clay.

Quantity: estimates are of 250,000 or more tonnes of limestone in the Muhokya areas, where there are several small-scale lime producers.

Quality and variation: marly and light grey, the limestone is good and suitable for the production of lime.

The limestone for lime production is obtained from a quarry near the kiln and the deposit is either white or grey in colour. At current levels of production the reserves will last 10-15 years. The white limestone is a bit yellowish but is preferred by customers for white-washing though it contains a lot of siliceous and iron oxides. The grey stones are denser and require a lot of energy input for all steps of processing. Due to a shortage of drilling equipment, most of the limestone is mined from the quarry manually using hammers, hoes, pick axes and crow bars after overburden removal using hoes. Overburden is not great as there is little soil due to soil erosion and plant growth is stunted by the hard limestone rock.

The limestone from the quarry is transported to the crushing site, 300 metres away by a tractor with a hydraulically operated tipping wagon. The tractor runs on diesel.

Production technology and processes

After transportation to the kiln site, the big sized stones (approx. 10-20 kg) are crushed to egg/fist sizes (about 80 mm). This crushing is carried out manually, by 3-6 men on contract basis and they produce about 9 wheelbarrows per day (about 65 kg). These stones are then loaded into the kiln via an access ramp. The kiln is loaded with 8-12 courses (one course contains 200 kg of firewood and 10 wheelbarrows 650 kg of limestone. A lorry load of firewood (5-6 tonnes) costs Sh 200,000 ($200). The firewood is weighed on an improvised balance and loading is alternate: firewood first, then limestone and so on. Firewood is obtained from eucalyptus plantations some as far away as 50 kms and during the rainy seasons this creates problems. The firing is continuous and production is 4-5 tonnes per day of quick lime. After firing, the limestone is off-loaded using wheelbarrows: 120 wheelbarrows per day are transported to the works shed, about 50 metres away.

Here the quicklime is hydrated manually using a hosepipe. After 48 hours, the slaked lime is sieved in various sizes: 0-1 mm falls to the packing area, 0-2 mm is ball milled and the resulting 0-1 mm joins that on the packing area.

Due to present low demand, burning is carried out about one week every 2 months. If a contract is received to supply, for example, lime for a road-making project, Equator Lime is capable of increasing production up to 18 courses, which would then be the maximum capacity of the kiln. The sieving is done by an electrically driven rotating sieve. The electricity is provided by a diesel generator (modified tractor) of 11 KVA, using 2.5 litres of diesel per tonne. Here coarse particles and lime are also separated. The oversize material is transported by a conveyor belt into a $3m^3$ metal container. This material is then collected daily by a tipper for refiring in the kiln.

The lime is stored in a big shed and 2-3 people pack it manually into 25 kg bags. One man can pack up to 150 bags per day. The weight of each bag is spot-checked by a foreman.

The packed lime is then stored in a loading bay, which is open on the sides (no walls) and this causes problems of wetness in the rainy season. From the storage/loading bay, customers come to collect the lime and the company also transports the lime to customers in various towns and trading centres, including centres in Zaire.

For quality control, the company submits samples of limestone and hydrated lime regularly to the Department of Geological Survey and Mines for chemical and physical tests. At the same time, the company has its own test-sieves for the testing of lime fineness. These tests are also cross-checked with other laboratories in the country.

At the time of writing, the grey lime was costing Sh 3,000 ($3) per 25 kg, while the white lime cost Sh 3,000 ($3) per 25 kg bag.

Investment

The design of the whole lime production process including the buildings and the machinery were all done and built by the company's workshop. About $40,000 has been invested in the lime production though the company still operates at a loss. The plant is owned by: Expo Mechanical (Kampala), Wolfgang Schulmester (Kasese) and Dominik Schmidlin (Kasese). The last two are operating and running the plant in Kasese, while Expo Mechanical takes advantage of its location in Kampala to facilitate the marketing of the lime. All the investment came out of their private funding.

Table 35: Equator Lime's production of lime

Year	Quantity (tonnes)
1978	101
1979	30
1980	76
1981	84
1982	74
1983	413
1984	165.1
1985	120.4
1986	10.7
1987	14.5
1988	210.5
1989	193.4
1990	74.0
1991	1770.0
1993	210.7
TOTAL	**3,446.1**

The majority of the workers are unskilled and some have to obtain on-the-job training. Absenteeism is a major production problem.

The workers are paid Sh 800 ($0.8) per 8-hour day for non-skilled workers and Sh 1,200 ($1.20) per day for skilled workers. For early shifts, Sh 400 is given as extra bonus. For the quarry workers, a contract system is used.

Production Costs

To produce 1 tonne of hydrated lime costs Sh 66,000 ($66) split into:

Wages	20%
Fuel	50%
Machines	20%
Bagging	10%

The finished lime is then sold at Sh 80,000 per tonne ($ 80).

Waste: Due to various unfavourable production constraints, there is a cost in production due to small spoils (5%) and about 10% due to milling (sand and clay). Inefficient screening also creates other wastes.

Markets

The major markets are in road construction and building construction, though diversification into lime pozzolana promises another channel for lime consumption. Road construction, though the biggest consumer, tends to be seasonal in that their lime requirements depend on availability of contracts to build roads. At the same time, these road contractors require large quantities of lime per day and per kilometre (120 tonnes of lime per km) and to be sure of steady supply they usually instead sign a contract with big lime producers. Equator Lime's production is small and chances of supplying a major consumer are limited. Equator Lime has to sometimes team up with other lime producers in order to be able to supply to the big road contractors.

Customers come from near and far to buy grey lime, white lime and now pozzolana cement. They collect these themselves or buy it in urban centre shops. There is no detailed data on who takes what, where.

Policies and regulations

Usually before one goes into lime production, one has to obtain a prospecting license which leads to a mining license. These enable a person to extract the limestone for lime production. At present a prospecting license costs Sh 28,000 (Sh 20,000 deposit and Sh 8,000 annual fee). A mining fee in the form of royalty is paid depending on the quantity of limestone extracted.

Most of the fuel used in lime production is wood, purchased from wood plantations, both private and Government. One can also purchase a wood plantation but will have to pay extra fees in the forms of a plantation fee and local taxes. With the present decline in wood supply due to deforestation, lime producers are being encouraged to grow their own wood lots. Equator Lime is already doing this to safeguard its future.

Because of the unhealthy conditions in the quarry and plant, the lime producers are being encouraged to provide health and safety tools and equipment (e.g. gas masks, boots, coats, etc.,) to protect the workers. There are also provisions and regulations on the reclaiming of quarries from where the limestone is extracted. These must be refilled or reclaimed so that no water can stagnate there and be a source of diseases like malaria.

Constraints to production and marketing

Due to poor kiln design, high productivity and good quality lime are not achieved. Raw materials for kiln construction like refractory bricks and insulators are not available locally and when imported are costly. Because of relatively poor lime produced, Equator Lime finds it difficult to sell.

Equator Lime is presently finding it difficult to operate on its present site due to the high rent they pay the owner of the land. They are now looking for their own site where they intend to construct a 20-tonnes-per-day kiln which will be fired with agricultural waste materials.

Future plans
In future the company has plans to increase lime pozzolana cement production after they have expanded their lime production facilities. They have an improvised grinding machine, supplemented by a ball mill to grind to the desired fineness. But they are now in the process of acquiring a bigger ball mill to cater for increased fineness and expanded production.

Hima Lime Works, Uganda

W. Balu-Tabaaro

The Hima Lime Works is 19km from Kasese town on Kasese-Fort Portal road. It is about 250m from the main road and is next to the Uganda Cement Industry-Hima.

The limestone deposit was discovered at Hima, Kasese in 1943. Kilembe Mines Ltd, then the only major industry in the locality was later granted a license to exploit the limestone deposits for copper processing. The mine leased part of the limestone deposits and set up a small-scale lime producing plant known as Hima Lime Works purposely for copper processing. Thus Hima Lime Works belongs to Kilembe Mine Ltd. When copper processing ceased, the plant continued to produce slaked lime both for building and road construction. The production of lime was boosted when the Ministry of Finance loaned some funds to the Lime Works in 1988 and 1990. With these loans the Lime Works acquired a tipper lorry, a tractor, Dyna pick-up and modified the plant to increase production. The lime works was also able to supply the individual customers and some road contractors. The loans were repaid.

Presently, the lime is being produced for building and road construction. Coarse lime for agricultural use is now available at the site. Samples of lime for sugar production and animal feeds have been taken by research stations and institutions. In future, the plant hopes to supply other industries such as chemical industries (e.g soap industry), the pharmaceutical industry, sugar industry, leather and tanning industry and water treatment plants. The lime works sometimes supplies the limestone to steels works such as Uganda Steel Rolling Mills for the extraction of iron from ores or scrap.

The total Hima Lime Works' land area is 243.80 ha and the currently used area of 105.2ha is as follows:

Quarry (being utilized)	34.7 ha
Houses	3.0 ha
Eucalyptus trees	67.5 ha

The rest of the area is reserve limestone deposits and gardens.

Raw materials

The limestone is a sedimentary type of limestone. The calcium carbonate content is high.

The limestone deposits are generally compact, and in three flat lying strata of thickness up to about 1 metre. These strata are separated by layers of alluvial sand of uneven thickness and the whole is overlain by an overburden of weathered materials.

The estimated reserves of limestone are 70 million tonnes.

Mining of limestone

The limestone is extracted by open-cast system. It involves removal of the waste materials, drilling and blasting. The weathered overburden is primarily removed using a caterpillar excavator and the little which remains is cleaned off manually before drilling. The drilling is done using compressed air and the drilling machine is operated manually.

The holes drilled are then blasted with explosives. The limestone deposit being mined is about 200m from the production site. It is transported to the kilns for burning using tramming cars. The lorry and tractor use diesel while the cars need electrical power and manpower (when there is no electrical power).

Production technology and processes

Size reduction: stone less than 125mm across is taken directly to the kiln for burning. Stone larger than this is first taken to a jaw crusher for size reduction. Rocks which cannot be loaded onto the tipper truck are broken down manually using hammers. manually loaded into a tipper lorry or the trailer of a tractor which is then off-loaded near the conveyor of the crusher ready for crushing. The stones which are too big and heavy to load into the lorry or the trailer are first reduced in size manually using 14 lb and 7 lb hammers.

Loading the kilns: there are 5 small vertical kilns originally each 8 ft diameter and 13 ft effective height, with a suspended hearth 1.50 m above lower ground level built into an embankment. The bottoms of the kilns are lined with steel rods to support the fuel and limestone.

To load a kiln, alternating layers of fuel (firewood) and limestone are loaded into the kiln until it is filled up.

Calcining (burning) of limestone: the loaded kilns are fired from below and the mixture of fuel and limestone burns upwards. If the level of the burning mixture drops, more fuel and limestone are added until the kilns are full. It takes an average of 4 days when using firewood (or 6 days when using charcoal) to produce a kiln of quicklime ready for off-loading.

The charging and burning of limestone is a batch process.

The quicklime is off-loaded into bigger jubilee wagons each of capacity 0.92 tonnes by shaking the steel rods at the bottom of the kilns.

Transportation of quicklime to shed: the loaded wagons are pushed manually on the narrow gauge rails to the limeshed. The quicklime is tipped off on one side of the track in the shed by manually pushing one side of the wagons.

Slaking (hydrating) of lime: the slaking is done manually by spraying water on the quicklime. An exothermic chemical reaction occurs with much heat given off producing steam and crumbling of quicklime into a powder of slaked lime which ends up dry if the quantity of water is controlled.

Slaking is done by experienced workers who know how to control the amount of water to be added.

Screening of lime: The slaked lime is screened manually in two stages: The first stage is primary screening using 6 - 8mm sieves to remove the unburnt limestone (rejects) which is recycled for further burning. The second stage is secondary screening when using 1 - 1.5mm sieves to obtain fine lime and coarse lime (agricultural lime).

Packing of lime: the bagging of lime is done manually. The fine lime (all of which passed through a 1.468 mm sieve) is bagged in 20 kg or 25 kg 3-fold paper bags using a bagging funnel. The coarse lime (agricultural lime) is stockpiled or bagged in 50 kg plastic bags.

Plant modification in the kilns and shed

The slow processes make the production low. When the loan was granted, it was proposed that the plant should be modified to increase the production.

In the shed, it was proposed that the quicklime should be crushed, milled, slaked, and screened before bagging. Thus the following were installed and are in position: Conveyer, jaw crusher, hammer mill, two screw conveyers, slaking tank and double vibrating screen. A fan was installed near the hammer mill to suck off the dust produced by the mill. Also electrical installations were done. The system was made to work but due to the presence of unburnt limestone in the quicklime, the quality of the product (slaked lime) was found to be lower than that of the product obtained previously using fewer processing stages. The kilns are short and thus not all limestone gets burnt properly. Hence the old system is still being used.

Equipment and structures

The machines/equipment and structures at Hima Lime Works are old. They include:

Air compressor: a Holman Compressor which is old but still operational.

Air receivers: they are two and made of steel. They are old and rusty and hence air leakage can easily be caused.

Air pipes: (hose and metallic)- The hose pipes are damaged and the metallic pipes are old.

Jaw crushers: these are old Allis jaw crushers with liners worn out and requiring new bearings. One is in the quarry and another in the limeshed.

Conveyors: they are old and need new rollers, idlers and belts. They have been refurbished.

Tramming cars: (Jubilee wagons) - They are worn out especially at the bottoms.

Winch machine and wire rope: the winch machine is old. Its wire rope is worn out.

Motors: the motors are old.

Kilns: each is of diameter 2.40 m and height 3.90 m. They have been rehabilitated.

Hammer mills: fabricated in a machine shop

Screw conveyors: the two were refurbished and installed.

Lime shed: the front and the sides are open. The roof and the rear are made of old iron sheets. The floor is concreted.

Tipper lorry UXV 280: it is in working condition but has mechanical problems and requires new tyres.

Tractor UXV: it needs new tyres and maintenance. It is in working condition. Its trailers are old and worn out and they need tyres also.

Dyna Pick-up UPI 294: it requires mechanical maintenance and new tyres. It is in working condition.

Office/store block: the walls are permanent and have cracks caused by heavy vibrating as a result of nearby blasting.

Houses in camp: the houses are single room houses. The walls and floors have cracks and the roofs are made of iron sheets.

Source of water and energy used

The source of water used at the lime works is Mubuku river which is very reliable. The water is used both for human consumption and slaking lime. The water supplied to the lime works is always treated. There is a bore-hole near the shed having very salty water which cannot be used for drinking nor washing. The water can be used for slaking lime in case there is a mechanical problem with the water pump at Mubuku river or pipe leakages.

Initially wood charcoal was used to produce lime but later abandoned due to the difficulty in getting it. Currently (since March 1990), firewood is being used to produce lime. Supplied from Fort Portal (80 km from the site) or Nkombe Saw Mill, Bunyaruguru (Bushenyi), it comes as big logs of diameters 15-60 cm of hard and matured wood.

A lorry load of fuelwood is purchased at USh 120,000 (US$120) delivered to site: 7.2-9.6 m^3 of firewood is needed to produce one tonne of lime.

Quality of products made and quality control

The products of lime processing are fine lime and coarse lime. Fine lime is that of which 100% has passed through 1.168 mm sieves and an average of 95% has passed through 0.425 mm sieves. It has over 55% CaO and 70% Ca(OH)$_2$. Hence the hydrated calcium lime complies with all the requirements of relevant standards.

The laboratory for quality control is at Kilembe (not at the site) but lacks some chemicals and apparatus.

Total production capacity and actual output

When the kilns were modified to increase their capacity it rose from 200 tonnes/month to 440 tonnes per month. Otherwise with the modification in the lime shed working, the total capacity would be 700 tonnes per month.

Primitive methods of slaking, screening and packing, woodfuel and marketing constraints have constantly kept the production figures very low. The actual output per month is 60-320 tonnes.

Table 36: Hima Lime Works' production of lime (tonnes)

Year	Quantity
1961	1,716
1966	1,750
1970	1,261
1971	1,410
1972	1,240
1973	16,200
1974	14,300
1975	9,600
1976	614
1979	17,724
1980	528
1981	900
1982	22,321
1983	21,882
1984	19,168
1985	7,482
1986	14,962
1987	10,723
1988	16,750
1998	5,112
1990	16,760
TOTAL	**240,403**

Labour

At present there are 34 employees and 10 casual labourers as follows:
Administration office - 7, quarry section - 8, kiln section - 11, limeshed section - 13, sanitary - 1, and security - 4.

Most of the workers are not skilled and thus need training. The 34 full-time employees work 8 hours per shift all year around. The 10 casual labourers are employed every three months. The full-time employees are paid according to their grades and per shift they have worked in a month. Those who bag lime are paid Sh 25 for every bag of lime packed.

Waste

Waste in processing the stone for burning is considered to be small, but significant quantities of waste are produced in the burning and subsequent processing stages. A typical burn produces 17 tonnes of lime and 4.1 tonnes rejects. The rejects are waste obtained during screening. Lime dust waste is also produced. Additionally, some bags get torn during handling and transportation.

The overburden soil obtained during quarrying can be used to refill the parts of the quarry which have already been mined and contain water. It can also be used to level the roads in the quarry. The rejects are recycled for further burning and are sometimes used to refill the potholes in the camp and roads. The damaged lime is re-screened and bagged or it is used to white wash the buildings in the camp. The lime dust is trapped and bagged when the fan is used. Otherwise when screening it is difficult to trap the dust.

Table 37: Hima Lime Works' production costs (per month in USh)

Salaries and wages	650,000
Fuel (firewood)	3,360,000
Fuel and oils	400,000
Travel and subsistence	200,000
Screening and bagging	150,000
Loading and off-loading	60,000
Lease, rent & licence	106,000
House rents	18,000
Electricity	500,000
Lime bags	2,800,000
Replacement/ maintenance	500,000
Tax	300,000
Committee and lunch	20,000
Newspapers	14,000
Explosives	1,500,000
Miscellaneous	1,100,000
TOTAL	**11,768,000**

Markets

The lime market is very volatile and yet very tight. It is these marketing constraints that have constantly kept production figures low. The biggest customers are normally road contractors and thus the major markets are not really close.

The selling price of lime has been erratic. The Uganda Lime Producers Association formed in December 1989 has since set control prices for all lime producers and co-ordinates lime production/supply programmes with consumers, mainly road contractors. The current price for contractors is $ 86.68 per tonne of bagged hydrated lime.

Policies and regulations

The licence for raw materials is obtained from the Geological Survey and Mines Department, Entebbe.

The assessment rates, ground rates, operation permit and limestone quarry permit are paid to Hima Town Board.

No licence is paid for the woodfuel used because

the plant pays for the fuel delivered at the site by suppliers.

The price of lime is controlled by the Uganda Lime Producers Association.

The price of the raw material (limestone) and wood fuel delivered are set by the Management of Kilembe Mines Ltd.

The wages are paid according to Kilembe Mines Ltd regulations regarding wages.

For health and safety, the following are provided to workers; masks, boots, overalls, elements, etc. There is a dispensary at the site to attend to sick workers and their families.

The lime dust produced during milling is sucked by a fan.

Any contractors who need big quantities of lime have to put in an order, and enter into an agreement which specifies that when paying, it should be done through the Bank.

Main bottlenecks and problems

- Scarcity of wood fuel
- inefficiency in drilling and loading limestone into the lorry or trailer of a tractor.
- kiln inefficiency and constant crusher breakdown.
- purchasing of explosives.
- inefficiency in screening and bagging.
- small labour force.
- purchasing paper bags.
- no ready major markets available.

Plans and prospects

- Purchase of modern drilling machine and a wheel loader.
- purchase of new liners for the crusher or purchase new crushers
- construct more vertical kilns or purchase preferably a horizontal rotary kiln expected to use heavy furnace oil.
- contact a firm which can efficiently and promptly supply explosives whenever need arises.
- recruit more labour so that at least two shifts per day could be run.
- purchase the packing plant
- purchase vehicles to ease transport
- complete the modifications proposed in the plant so that the targeted tonnage is reached.

Homa Lime Company, Kenya

J.P. Brooks

Raw materials

The limestone deposits in Koru are carbonatite of volcanic origin. There are an estimated 65 million tonnes of raw materials with an annual consumption of about 30,000 tonnes. The stone consists of various levels of purity with an average content of 70% $CaCO_3$. Impurities include compounds of phosphorus, averaging 2%, and also manganese, iron and large quantities of soils, on the surface of the workings.

At present the rock is blasted using ANFO and a Cat D7 is used to push the stone. The stone is either broken and sorted by hand or processed in a crushing/screening plant with capacity of about 50 tonnes/hour. Hand quarrying produces between 2 and 3 tonnes of clean stone/day.

The distance from the quarry to the production site is 3 km, and to move to 70 tonnes of stone/8 hour shift requires one 60kW agricultural tractor and 8 tonne tipping trailer unit.

Production technology and processes

Crushing and screening plant: comprises of a grizzly feeder into the primary crusher, jaw size is 60 x 75 cm, powered by a 75 kW motor. The product from this is conveyed to a vibrating screen with 2 mesh sizes, i.e. >50mm, 2-50mm and the rest <2mm. The first and last fractions are removed to stockpiles by conveyor. The middle fractions falls into a secondary crusher, also powered by a 75 kW. motor. The resulting gravel passes through another vibrating screen to give three sizes, >4mm, 2-4mm and <2mm.

Kiln: the vertical shaft is 11m high. It is constructed of an outer casing of concrete insulated on the inside with diatomite. Inside the diatomite is a lining of insert stone such as sandstone which is in turn protected by refractory bricks cemented to the sandstone layer using fireclay. The kiln works on a mixed feed basis with alternate layers of wood and stone. The stone and wood is loaded manually and burnt in the kiln at temperatures in excess of 960°C. The quicklime is drawn from the bottom of the kiln through four doors using wheelbarrows.

Hydration plant: the burnt stone is fed into a crusher to reduce the size to facilitate hydration. The ground product is then elevated to a storage hopper from where it is fed onto a table feeder. This is a revolving disc which controls the amount of quicklime that enters the premixer where a small amount of water is added to mix with the quicklime. The hydration reaction is exothermic (i.e. heat generating) which dries off the slaked lime product. Steam from the reaction is drawn out to the atmosphere via a scrubber where lime dust is reclaimed back into the system. The hydrated lime is elevated into air separators where the lime is separated from the improperly burnt particles which have not hydrated. The hydrate is bagged while the tailings are removed for other purposes. Bagging is done in 25 kg paper sacks.

Energy requirements: energy is obtained from three sources in the production cycle. Fossil fuels provided the energy to drive the transport for moving the raw material and the finished product. The energy for calcination is obtained from woodfuel mostly from eucalyptus plantations maintained both on the company's property and externally. This wood has a calorific value of approx. 24,000 KJ/kg. Some indigenous wood may also be used with a calorific value of about 60% of the above.

The crushing/screening plant and the hydration plant are dependent on electrical energy - approx. 18 kWh per tonne of calcium hydroxide.

Quality of products: the quality of hydrate is ensured by collecting random samples which are tested for particle size and available calcium oxide content in the company's laboratory. More detailed tests are carried out on our behalf by organizations such as the Kenya Bureau of Standards and Kenya Industrial Research and Development Institute. The lime is manufactured to confirm to International Standards such as BSS 890/1972.

The design of the production line was originally obtained from overseas but most of the plant was actually fabricated in Kenya and installed using the services of both professional engineers and company staff.

Investment

The plant and land on which the limestone deposits occur is owned by Homa Lime Co. Ltd. The funding came from the shareholders and from revenue generated from sales.

Labour

The total average labour force directly involved in the production process is 81. Except for jobs like blasting which require a licence, qualifications are not required except those obtained through experience

Skill shortcomings are not a problem as operators are taught on the job wherever possible. Both permanent and temporary workers are employed depending on demand for the product. Unskilled

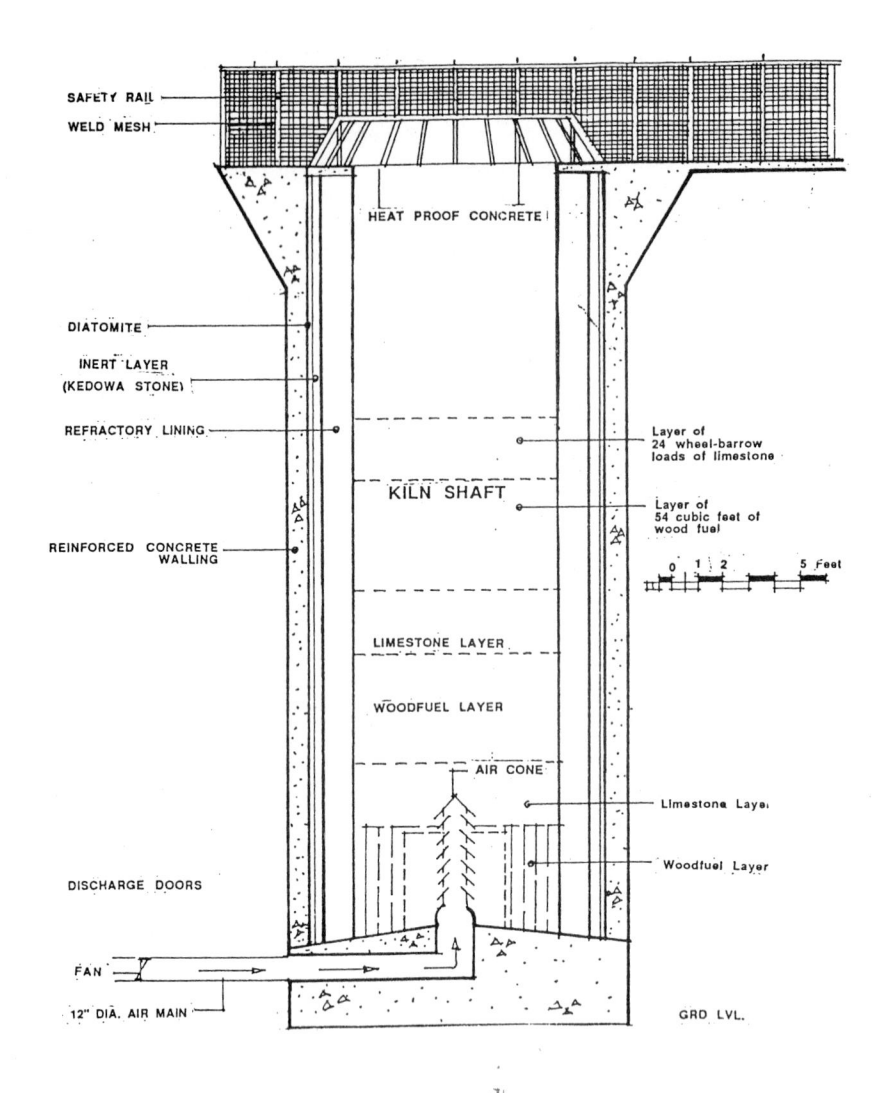

Figure 3: Structural diagram of kiln at Homa Lime's factory in Kenya

jobs such as quarrying and loading are paid for on a piece rate basis.

Operations continue throughout the year when possible but this is dependent on demand. Those people who do not work on a piece rate basis work on an 8-hour shift.

Casual and piece rate labour are paid fortnightly whilst permanent employees are paid monthly. Rates paid vary according to skills, length of service and collective agreement with the union.

Production costs	%
Wages	9
Quarry	14
Woodfuel/Electricity	20
Packing Materials	17
Transport	9
Depreciation	9
Overheads	22

The average cost per tonne is KSh 2,854 (approx. US$60).

Waste

Waste in quarrying, crushing and screening: between 50-70% is useless for burning in our kilns due to small size or contamination with soil. However, about 60% of the waste can be utilized for other purposes such as agricultural lime, stockfeed lime by the animal feed millers. Other quarry by-products can be used for chicken grit, road gravelling material, filler for building sites and agricultural lime. One is a coarse powder comprising a mixture of calcium carbonate, calcium oxide and calcium hydroxide. This is sold as an agricultural lime and is more reactive in the soil due to its higher proportion of the carbonate present. This product can also be used as a soil conditioner and it also an excellent material for gravelling unsealed roads.

Markets

Product price per tonne	KSh ($)
Calcium hydroxide	4,200 (90)
Calcium Carbonate (stockfeed)	845
By-product (agricultural)	500
Limestone	450

Marketing efforts include visiting customers and potential users, advertising, sending out circular and conducting demonstrations.

Sales per month	tonnes
Hydrate	
chemical industry	600
building industry	100
road construction	200
Stockfeed for animal feeds	100
By-product for agriculture	200

Sales throughout the year are fairly constant except during long rains in April/May when the sugar millers usually close for annual maintenance.

Sales history for hydrate	tonnes
1991	11,700
1992	6,327
1993	8,806

The reduction in sales between 1991 and 1992 was due to the almost complete absence of road construction. This is however, expected to improve over the next year or two.

Policies and regulations

Licences are obtained annually as follows :
   Miscellaneous occupation licence–export
   Manufacturer's licence

Pricing regulations: there are no controls except those imposed by market forces.

Wage regulations: we follow guidelines imposed by the Ministry of Labour and also have a Collective Agreement with a Union.

Health and safety regulations: protective clothing where appropriate is supplied. Employees are also

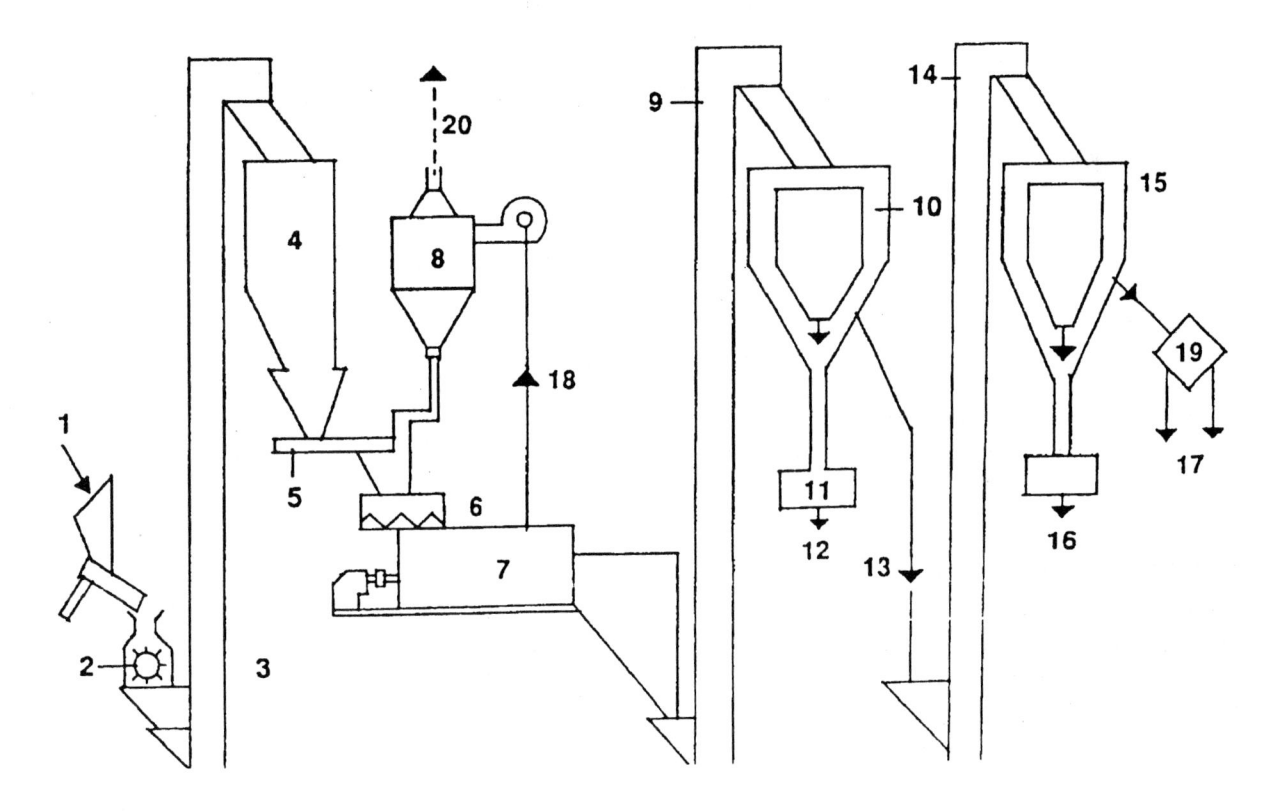

Flow plan of the hydration plant
1. Quicklime feed hopper
2. Crusher
3. Elevator
4. Crushed lime storage hopper
5. Table feeder
6. Pre-mixer
7. Hydrator
8. Washer/scrubber
9. Elevator
10. Air separator (primary)
11. Bagging machine
12. Hydrate of lime
13. Tailings/outlet
14. Elevator
15. Air separator (secondary)
16. Hydrate of lime
17. Tailings
18. Hot milk of lime to washer/scrubber
19. Screen
20. Scrubbed exhaust

Figure 4: Flow plan of the hydration plant at Homa Lime's factory in Kenya

A re-used quarry brings wide environmental benefits in Kenya

examined medically by medical officers from the Health Department. Normal safety procedures and testing of machinery are carried out as required by the relevant authorities.

Environmental regulations: such regulations that exist are complied with. The most obvious pollution is the smoke from the kilns. It is in our interest to minimize this for reasons of fuel efficiency. The nature of the fuel however means that the content of toxic wastes is low. Re-use of quarries is not compulsory but is possible in our case because of the quick regeneration of vegetation. They may then be used for grazing or for planting trees.

Main bottlenecks and problems.

Poor communications: e.g. telephones and related services are frequently inoperative.

Transport: expensive road transport rates and very inefficient rail systems result in expensive products to the end user.

Power supplies frequently break down.

Taxes: a great number of taxes are imposed on the manufacturer, employer and employee in the form of licences, cess and service charges. Despite all these taxes, services supplied are minimal and often non existent resulting in businesses having to supply them and results in increased costs to the consumer.

In addition, there is the almost universal inefficiency and corruption that exists in many offices, notably in the public sector.

Plans and prospects

There is an intention to improve the screening of the quarry products so that we can produce building blocks and provide cleaner products for sale. The production line will be maintained as necessary and any improvements required to reduce waste or improve efficiency of production would be carried out.

The question of improved energy utilization, particularly that of wood consumption in the kiln, has been taken up with various experts. As yet no solution has been found.

Expansion of the plant in the near future is not envisaged as there is already unutilized excess capacity. At present only one kiln out of out of three is functional. It is essential therefore for us to try to develop a bigger market in the building industry.

Kenya Calcium Products, Tiwi, Kenya

I. Leckie

Raw materials

We have 500 acres of calcium carbonate limestone to an average depth of 18m. Currently we quarry 300 tonnes per day, therefore reserves are 'unlimited'. The stone is of a good quality at about 93% calcium carbonate content. There are harder layers which require more fuel and patches with higher iron content. However generally the stone is uniform and suitable for lime manufacture.

Extraction is generally by rip/dooze technique using a D8K Caterpillar excavator. If necessary we can blast, although this is expensive per tonne of rock obtained. Rip/Dooze suits our needs as all sizes of stone down to powder are utilized. Doozed material is machine loaded into trucks or tractor trailer units to be taken to the crusher, from within a 1 km radius. Should there be major breakdowns we can quickly switch to hand breaking and loading which for the same tonnage of kiln size would require 75 extra men plus double the transport.

From the crusher the kiln size (6-15 cm) stone is transported by tractor trailer to the kiln about 1.2 km away.

Production technology and process

The crusher is designed to process 100 tonnes per hour, but with maintenance time plus other stoppages the average is 50T per hour. The diagram, Figure 1, shows the grades of crushed and ground limestone produced and their principal uses.

Kiln size (7.5-15 cm, but can also mix 5-6 cm) stone is transported to kilns by tractor trailer. The kilns are vertical shafts with a height of 9.6 to 10.8m and a diameter of 2.4m. They are run as a continuous process operated on 3 x 8 hour shifts. The fuel used is currently woodfurl and the stone and wood are stacked in the kiln on clear alternate layers. The ratio used is approximately 2.5m^3 wood to make one tonne of 70% CaO content lime plus 0.3T lime waste.

The kiln is continuously fed, air is blown into the bottom for 4 hours then quicklime is drawn (120 wheelbarrows) for one hour and the cycle resumed. The factory is old, having been bought second-hand from Ireland but has the capacity of 60 tonnes hydrate plus 20 tonnes by-product per 24 hours. Figure 2 shows the lime production process.

The manpower required is about 50 at the lime production site over a 24 hour period. This number is doubled if taking woodfuel from store to kiln and lime from factory to store. The water is from our own well and is very brackish but this does not affect the quality of lime. With the erratic market of the last three years actual output has fallen from 60% of capacity to about 40%. This we plan to increase to 75% in the next two years by penetrating into the building market, an area that in the past we have not been able to explore.

Investment

The plant is owned and run by Kenyans. The initial capital came from the sister company Homa Lime; since then it has been run and maintained from income.

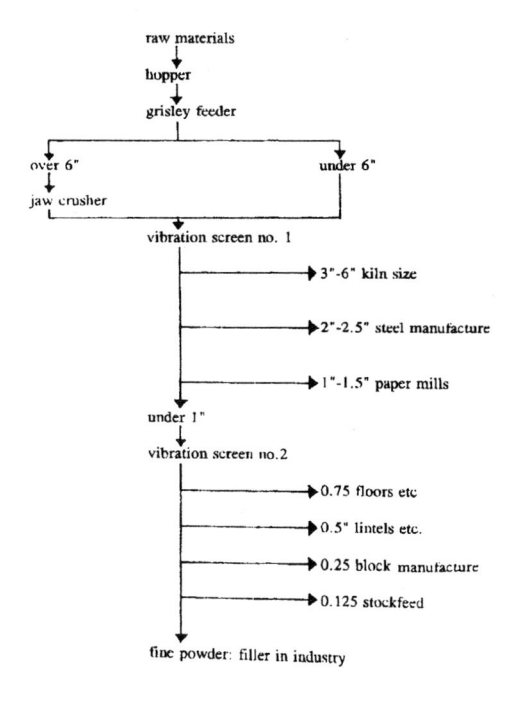

Figure 5: Use of production from Kenya Calcium

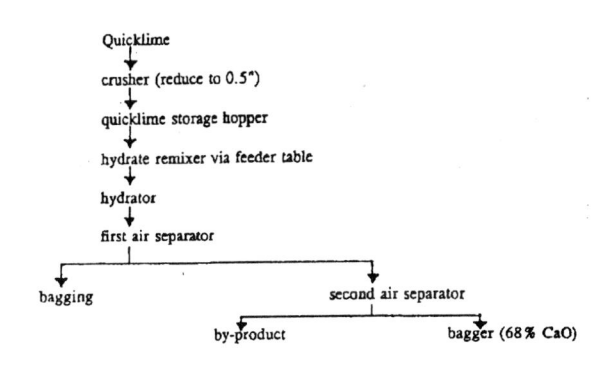

Figure 6: Kenya Calcium's lime production process

Labour

Job	Number	Wages (KSh/mo)
Kiln	15	27,000
Kiln drawers	7	13,650
Lime processors	5	9,000
Mechanics	4	9,000
Machine attendants	3	6,600
Fuel movement	9	16,200
Supervisors	3	6,000
Drivers tractor	8	16,000
Drivers lorry	8	69,000
Drivers wheel loader	3	69,000
Workshop	8	24,000
Driver bulldozer	2	5,200
Quarry blaster	1	2,800
Quarry compressor	4	6,000
Security	22	25,000
Office staff	10	100,000
TOTAL	112	404,450
		($8,600)

plus others brought in as required.

Production costs

a) Variable costs

Quarry	20%
Fuel	32%
Production	20%
Bags	25%
Handling	3%
Total	100%

b) Total costs

Variable costs	60%
Overheads costs	40%

Waste

As we have markets for all sizes of stone either as aggregate or powder there is a theoretical zero waste. In the factory due to old fashioned kilns giving poor burning there is 30% waste. This is stored outside and will eventually be sold as a fertiliser and/or a cheap building product.

Markets

Sales are oriented towards sugar factories and, occasionally, roads. Now that we have a definite price advantage over cement, a third and perhaps the most important market is added: lime as a cheap building material.

As with all small-scale lime burners a road project pushes production to the limit and the absence of a road project results in half production and full stores. Therefore it is vital to get into the building market to give a more consistent sales level.

Should capital be available at affordable rates a vigorous marketing campaign in the building and construction Industry will ensure a steady income. Without this lime sales will continue to depend on aid road projects and the sugar industry.

Policy and regulations

These are many and varied in Kenya at both local and national level. If they could be somehow combined much confusion would be eliminated. Minimum wage levels are set by the Government but we follow our Union negotiated Agreements which are higher.

Problem areas

Relying on income to maintain and replace machinery is no longer realistic unless new markets can be found. With all inputs rising at excessive levels and customer resistance in using lime as a building material machines have to be 'kept going'. Loan money in Kenya is so expensive that new machinery cannot be bought when required. The long term result will mean inefficient lime production.

One of the biggest problems for us is that having created a relatively cheap product the bulky material needs to be transported to the customer. Transport by road is extremely expensive. Transport by rail, while cheaper, is extremely unreliable. Kenya rail must improve for us to be able to market our lime fertiliser in particular, but the same applies to a lesser extent to all products.

Being situated on the south coast we have to use the ferry service to reach the railhead in Mombasa. This is expensive, subject to delays and results in a lorry only carrying 2 trips a day for a 30 km round trip. A causeway north of the creek will be most welcome for our transport logistics.

Kenya Power and Lighting are erratic and seem to have given up on the idea that 3 phases should each hold equal current. The result is long down times from power cuts and motor burnouts from fluctuations and phase failures. If a new plant with sophisticated machinery was in place, power supply would rank as one of the top three worries along with access to finance and the erratic market.

The future

The quarry by-products (chipping and powder) must all be sold and production increased. The Lime By-Product must be marketed as a fertilizer so that there are no waste elements. Lime must be sold as a cheap building material to create a more even sales pattern.

A more efficient fuel than wood must be found. We intend to convert before the end of 1994. This is not merely wishful thinking, but simple action required to survive as a viable business. With fairly fixed costs per tonne, the best way to generate income is to increase turnover.

Production of lime in Zanzibar

Jokha S. Suleiman

Lime demand in Zanzibar

Nearly all construction in Zanzibar uses lime. In conservation work of the many historic buildings lime plays a major role. It is also used in low cost roads in rural areas in the form of limestones aggregates and fillers.

It is somehow unfortunate for Zanzibar that new technologies have invaded the market and new building materials introduced (especially in the urban areas). Most of them are expensive and scarce due to import restrictions and non regular supply from the factories. The common example of these material is cement which to a large extent has substituted the use of lime. The price per square metre for structures using cement as the binder are normally four to five times that of lime constructions whereas there is little difference in the durability and life span (say 10-15%) based on the typical (maximum) four storey structures erected in Zanzibar.

The high price of cement is now a major problem for the development of the construction industry, most people cannot afford to build and so housing problems have increased. Thus the lime technology has again started to take over. Common examples of local building materials incorporating lime are:

- stabilized soil blocks (lime and laterite 1:3)
- cement, lime and laterite blocks (1:1:9)
- stone masonry walls

About 20,000 tonnes of lime have been estimated to be needed for the Unguja island per annum. Also the future proposals which were not included in the previous considerations are the possibilities of exporting lime and the need to look for other uses of lime rather than construction works. Therefore it is clear that lime was and still is one of the materials in great demand in Zanzibar.

Raw materials and fuel
Limestone

The surveyed area of Dimani, of about $720,000m^2$, has a reserve of twenty million tonnes of limestone. Up to now only about 2% of this deposit has been used. There are also other numerous deposits used for commercial lime production throughout the island. The reserves surveyed would be sufficient to meet the market demand of the island for 500 years, based on production of 20,000 tonnes of lime per annum.

None of the producers carry out any laboratory tests on their products. The local method of quality control is simply to check the softness of the quicklime. When water is added for slaking and the remaining limestone does not break down it is considered that the lime is not burned and so is of poor quality. In such cases, the unfired lumps are mixed with the next load of limestone in the kiln to be burnt. There are few users who are doing quicklime tests before use. One exception is the Mahonda sugar factory and the 'Kishore building contractors' who renovated the old dispensary of Zanzibar.

The extraction technology is simple. Limestones can be just picked up by hand on the ground surface or dug out using manual equipment, i.e. long pointed steel rods. This job is normally carried out by labourers who may or may not be among the members of the producers groups. Two labourers may collect one tonne per day if it is on the surface or half of that if it needs excavation. Lorries, cow carts and wheelbarrows are used for transporting the limestone to burning sites which are nearby, or an absolute maximum 8-10 kilometres away. The producers prefer to locate burning sites near to settlements to reduce transport distances to markets. This is a situation which is disturbing and could affect the long term health of the people living nearby.

Fuel

The fuel most commonly used for heap burning production is coconut logs. Coconut husks and shells are used as additives to increase the burning speed for both heap and the vertical shaft kilns. Currently, coconut logs are abundant but in the future, either other fuel sources will have to be sought or the harvesting and growing of coconut palm has to be organized on a forestry/rotation basis.

The Ministry of Trade Industry and Marketing of Zanzibar is among the government departments which are promoting small-scale industries such as lime production. They are conducting experiments on alternative fuel sources such as coconut husk blocks and rice husk blocks which are made from abundant resources currently regarded as waste. Samples of the test blocks were sent to other countries such as Malaysia, where these technologies have already been experienced.

Vertical shaft kilns on the island are fuelled by coal which is imported. The transport costs and the finite nature of the resource make it a less attractive alternative to timber. The fuel is normally dried out before use.

Production technology

Two methods of production are in use: heap burning

and vertical shaft kilns (VSK). The former is more commonly used because it is a simple technology with low capital requirement.

Equipment

The VSK method utilized more machine-oriented tools than heap burning. However due to lack of maintenance the machines are no longer working at the one functioning kiln at Dunga and have been replaced by manual methods. It can be said that, with the exception of the kiln itself, heap burning and VSK methods employ similar levels of mechanization and the equipment is as follows:

Type of equipment	Quantity
Wheelbarrows	6
Head pans*	10
Shovels	8
Sieves (1.18-1.2mm)	4
Buckets (20-litre)	6
Rock drills (4cmx150cm)	4
Plastic bags (50kg)	
Hammers (2kg approx weight)	10
Axes	4

* for adding limestone and as a main tool for measuring the ratio of materials

Production process

a) Vertical Shaft Kiln: As previously noted, the work is carried out manually. All the materials are thrown in at the top of the kiln. The task is repetitive though eased by forming the workmen into a human chain passing materials from one person to the next. For an output of 18 tonnes of quicklime, the process requires the following amount of material:

- 2,100 head pans (25.2 tonnes) of limestone
- 420 head pans (5 tonnes) of coal
- 3 sacks (50kg) of coconut shells
- 720 sacks (50kg) of coconut husks

The method itself involves the following steps:
a) Cleaning of bottom surfaces of the kiln
b) Starting the process by spreading 3 sacks of coconut shells over the bottom of the kiln; followed by:
c) 12 sacks of coconut husks then
d) 35 head pans of limestone and
e) 7 head pans of coal
f) c) to e) are repeated 59 times throughout the process
g) The kiln door is opened and the coconut shells set alight

Note: The kiln is full when materials are put in 25 times. The process is continuous by filling more materials when some of the lime is ready and removed at the bottom of the kiln.

About 30 labourers are involved and the full process takes about one week assuming all the raw materials are at the site.

b) Heap burning: For this method, limestone is burnt using firewood and coconut logs with the addition of coconut shells. Depending on the capital and space available, kilns are of differing sizes though all use the same ratio of materials. The following is the amount of material required for an output of 1.8 tonnes of quicklime:

- 2.5 tonnes of limestone
- 35 lengths (150mm x 300mm dia) of coconut logs
- 15 bundles of firewood
- 1 sack of coconut shells.

Method:
a) The firewood is arranged in a circular pattern
b) Coconut logs are placed on top of firewoods leaving holes for firepoints in both the firewood and coconut log layers.
c) Limestone is placed over the top and sides of the coconut logs, until all the logs and firewood are covered with the stones.
d) The coconut husks and firewoods are set alight via the fire points. The burning takes about 24 hours and employs two to three people.

For both methods tap or well water is used for slaking supplied by hose pipes, watering cans or buckets.

The lime technology, especially the heap burning method, was introduced in the island by the Malaysian people, Persians, Portuguese and Omanis (Ref.2-1991). The vertical shaft kiln is more recent, coming from the West and China.

Investment

Vertical Shaft Kiln (at Dunga)

Built in 1940s, by the local people with the supervision of a German citizen, it is now in a bad condition and the machinery no longer works. It was not possible to get the information of the capital cost. The present value including rehabilitation works is about TSh 50 million ($100,000) (rehabilitation would cost about TSh 30 million).

After the 1964 revolution, the plant was given to the National Service Department (JKU) and is now operated under a joint venture in partnership with private individuals. A contract was made between these two parties involving contributions for operating and maintenance costs. Such a kiln was built and operated a few times (3-4) and thereafter production stopped. It is thought that poor design or inefficient running were the source of the problems. In an attempt to get the kiln running effectively, the government is pursuing both assistance from donor agencies and the involvement of commercial interest in leasing the facility.

Heap burning

Below is a breakdown of the capital costs required

for producing 1.8 tonnes of quicklime using the heap burning method:

Equipment	Cost(TSh)
3 wheel barrows @ 8,000	24,000
Water tank (one)	8,000
3 water buckets @ 1,800	5,400
3 machetes @ 1,500	4,500
3 axes @ 3,000	9,000
3 hammers @ 1,200	3,600
3 shovels @ 1,200	3,600
Total	58,100

Materials
| Limestones | 7,500 |

Fuel
Coconut logs	7,500	
Firewoods and coconut shells	3,000	
Total		10,500

Labour
| 3 labourers @ 2,500 | 7,500 |

Overheads
| land compensation, etc. | 3,000 |

| TOTAL | 86,600 |

(Approx. TSh 90,000 equivalent to $180)

Normally the capital is from shareholders according to their agreed terms. The site for burning could be hired or sometimes may be a plot owned by one of the shareholders.

Labour

About 38 personnel are required to operate a VSK for an output of 18 tonnes within a period of one week continuously. This includes limestone collection and burning, making the assumption that all the required fuel is on site. With regards to the qualifications, at least two of the members who are the leaders should have been trained on the mainland to operate such kilns and they can then train others who at least have a minimum experience of heap burning.

Heap burning involves two to three people for a five tonne output in three days including burning and collection of materials. At least one of the group members should have experience of heap burning.

In the interviews we conducted it became apparent that the skill and practice is passed from generation to generation within families. The main problem facing the burners is the lack of know how in running a business (project management) as well as lack of quality control. Training is necessary in these fields. For the VSK workers, the employment is permanent and not only at times of lime burning, i.e. other jobs are carried out if the production stops. Incentives are given for good production figures. The heap burning workers employment is on periodical basis depending on the market demand.

Waste

There is only a small amount of waste of both limestone and fuel. Any unburnt logs or remaining pieces in cutting of wood can be dried and re-used as firewood. The remaining limestone dust can be used for block or brickmaking or as a floor filling (for making up levels).

Normally there is no waste in the end-product for either method. Unburnt limestone can be removed from the dry hydrated lime heap and added to the next batch for firing.

Production costs

Below is breakdown of total costs required for the two methods:

Vertical Shaft Kiln:
Output: 18 tonnes
Duration: one week assuming all the materials and fuel are at the site
Manpower: 36 labourers, 2 skilled heads

Wages (per day)	Cost
6 labourers @ TSh 250	10,500
2 heads @ TSh 300	4,200
20 helpers @ TSh 30 (from the Nat. services)	4,200
Raw materials	
Limestone 25 tonnes @ TSh 6,000	150,000
Energy	
Coal 25 tonnes @ TSh 34,000	85,000
Coconut husks 360 sacks @ TSh 300	108,000
Coconut shells 3 sacks @ TSh200	600
Overheads	
Equipment (approx)	10,000
Kiln (maintenance, security and storage)	10,000
Sub-total	382,500
Contingency 1%	3,825
TOTAL	**386,325**
Output 18 tonnes @ TSh 30,000	549,000

Profit = 28%

Heap burning:
Output = 1.8 tonnes
Duration = 2 days
Manpower = 3 personnel

a) Wages
2 labourers at TSh400 per day= 800.00
1 Head @ TSh 600 per day = 1,200.00
b) Raw materials
Limestone 2.5m^3= 7,500.00
c) Energy
Firewood 15 bundles= 3,750.00
Coconut shells = 500.00
Coconut logs= 6,250.00
d) Overheads
Equipment(approx)= 1,200.00
Total= TSh 21,200.00
Output = 72 bags @ 400TSh = 28,800.00
Profit@350 TSh23%

Market

The end-product (quicklime) is sold on average at a

cost of TSh 400-100 per packet of 30kg. The price variation depends upon the location (transport costs) and quality of the end product. The main difference between the two methods is that lime produced by the VSK method costs about 50% more than the heap method. Demand for lime becomes high at the peak construction period, i.e. when there is no rain. Currently, the demand for lime produced by the VSK method always outstrips supply. The situation is vice versa for heap burning where the production needs to be limited according to the orders requested.

It was not possible to get the sales figures due to lack of record keeping but it seems that in general, the production capacity has increased. The list of main purchasers of lime produced by the VSK method for the past three years as follows:
1. Stone Town Authority Office—Conservation Works
2. The Aga Khan Rehabilitation Project—The old Dispensary
3. UNESCO Training Programme and Rehabilitation Works—The Customs House
4. Mazrui Building Contractors—Conservation and New Works
5. Mr. Emerson (individual) Conservation and Development Works

The purchasers of the local heap burning method are normally private individuals and investors especially those constructing tourist hotels. We can see that all these are from the town. The lime is also used in large quantities in the rural areas for new construction but we can not trace these from the sales figures.

Policies and regulations

Permission to excavate or remove limestone and cutting of timber should be obtained from the Commission for Natural Resources. This falls under public land decree (Cap.93 - 1984 section 6(d)). This law is being re-drafted.

Until now, there has been no regulation regarding pricing. The market determines the price. This situation is causing problems for some burners who normally end up in loss or with low profit margins. The levels of the workers' wages is determined by their employer. If it is the government then the wages will depend upon the general set figures. for the local or private individual people this will depend upon the profit gained. Most of the times the private sector pays better wages. There are no health or safety regulations governing the burners. The burning process is normally undertaken without any safety precautions for the heap burners and few are provided at the VSK. At least the government burners are insured, but nothing of that sort is present in the private sector.

In Zanzibar, the Department of Environment has only recently been set up (five years ago). However, in the environmental policy (1992), there is a section which require any one who removes or cuts the natural products to obtain permission form the Commission for Natural Products. The opening of quarries could cause future problems by destroying the natural environment. Plans are to be made for their proper use over time and in each area. the Director of Environment has observed that the main problems associated with lime burning are the haphazard collection of raw materials and fuel. Also, though of lesser impact, the carbon-dioxide gas (produced in the burning process) is one among the greenhouse effect gases in the atmosphere. The people believe that the amount is not great but this is based solely on the present quantity produced without considering the previous (and future) amounts. Therefore the Department of Environment is considering the issue to make sure that the production of lime will not cause future environmental problems.

Main bottlenecks and problems

These fall under the following main groups:

Lack of sufficient funds: Although setting up costs are not that great for the heap burners nevertheless funding is a major problem. Inability to pay for raw materials and fuel results in low output and low returns. The problem is greater for the VSK. The dependence on imported coal from the mainland demands high initial capital.

Unstable market: The low quality of lime produced by the heap burning methods results in customer (especially those requiring large quantities) preference for lime produced by the VSK method. As a result piles of lime bags or lime heaps are often seen at the burning sites or by the road side awaiting purchase.

Poor quality control: For both methods no tests are carried out to check the strength of lime. It is assumed that the VSK method produces better quality lime resulting from the technology (whereby the burning temperature is more controllable).

Management skills: As previously noted, there is a lack of management skills to operate the business effectively in order to gain reasonable profit.

Modification (incentives): Lack of external support (such as free or low interest loans) means that any improvements are unlikely to occur. The business is at present not given a high priority in Zanzibar.

Information on lime: There is a lack of recording and no dissemination of information on production.

Plans and prospects

Outlined below are some of the proposals for the lime industry in Zanzibar. All are important and achievable given sufficient funds. The cause of delay in implementation is the lack of funds resulting from the country's poor economic situation.

a) To increase the production capacity and start

exporting lime. Feasibility studies are in process to check for its market.

b) Introduction of lime products other than for construction works, e.g. school chalks

c) Rehabilitation of the VSK at Dunga

d) Continuation of experiments and analysis of the STCDA demonstration lime kiln at Mazizini set up with ITDG experience.

e) Periodic seminars for educating the people in the proper ways of production and in the many uses of lime. Such a seminar was held on 27-28 Nov. 1991 under ILO and STCDA support.

f) To look for substitutes for coconut wood which could be scarce in the future and the destruction of which would damage the environment. To develop effective rotation harvesting of plantation timber.

g) Quality control of the procured lime. To educate the burners on the importance of checking the quality of lime produced. Utilizing existing laboratory faculties to carry out the necessary testing.

References:

Manufacture of Lime in Zanzibar - Industrial Development Unit-IDU/TAN/26 Commonwealth Fund for Technical co-operation, Marlborough House, London - September, 1985

Traditional and Current Uses of Lime Mortar - Fatma Kara - Stone Town Authority Zanzibar (report on the Mission to Zanzibar on behalf of ITDG (UK) for the Ministry of Finland Regional Programme for the Conservation of Cultural Heritage in SADCC countries, Rodney Melville and Partners 16 Nov - 2nd Dec. 1991

Zanzibar Stone Town Conservation Plan (summary of findings and draft planning proposals) MWCELE/AKTC - May 1993

Lime Seminar organized by The Stone Town Conservation and Development Authority, Rodney Melville and Partners, 16th Nov - 2nd Dec. 1991

Wachumico Co-operative Society, Tanzania

Huba Nguluma

The Wachoma Chokaa Minazikinda Co-operative Society (Wachumico) was registered in 1993. It is located at Kigamboni area along the Indian Ocean Seashore, almost 1km away from the Kigamboni Ferry, and about 2km from Dar es Salaam city centre.

The history of Wachumico lime production can be traced back to the period before colonialism in the 19th century. Lime was then produced using simple manual methods and used for aesthetic applications and building construction.

Each member of the co-operative now produces lime individually and looks for markets at his/her own cost. Wachumico as a society is responsible for organizing land acquisition, business licences, problem solving and represents its members in various fora. Members of the society contribute four bags of lime (25 kg each) per batch to the society. The bags are later sold and the money banked in the society's account for future development plans and immediate administration costs.

Raw Materials

Limestone: The limestone used by Wachumico to produce lime is procured from Sinda and Mnarani, about 25km and 10km away from the site respectively. Sinda and Mnarani are two small Islands in the Indian Ocean. The limestone obtained from the two Islands is said to be identical in quality. However, it was not possible to confirm this analytically. Traditionally limestone from the sea or ocean banks is called *matumbawe*. According to information available the *matumbawe* from the two islands are quarried manually by individuals, using crowbars, chisels and hammers. Wachumico purchases the *matumbawe* at a rate of TSh 6,000 (US$12) per one tonne or TSh 7,000 ($14) per 2 tonnes.

Fuelwood: Wood is the main source of fuel at Wachumico. Mango, coconut, and cashewnut trees are most commonly used. They are purchased from farmers at the rate of TSh 10,500 per 5-tonne truck and transported at TSh 9,500 per trip ($40 in total).

Water: Water for hydration is fetched directly from the ocean which is about 100m from the site. The largest clamp (kiln) can take up to 2,000 litres of water for hydration. This is a four to eight (4-8) hours' job for one or two person(s), respectively.

Production technology and processes

Lime production is based on traditional technology which involves:

Sizing of raw materials: For efficient kiln combustion, the raw materials have to be reduced to suitable sizes. Limestone is dressed to 10cm diameter while firewood is cut into 60cm length. One person dresses two tonnes of limestone within 1-2 hours.

Calcination: involves setting out, fire wood, piling of limestone and firing.

Setting out: the space for the clamp (kiln) is identified, cleared and levelled. Two common sizes are used with a radius of three or six arm lengths of a mature person or approximately one or two metres, then a circle is marked on the ground.

Firewood: is arranged in a circle in a single profile on the ground, the second row is laid perpendicularly towards the centre. The arrangement continues in this manner until the desired profile is made, usually rising about 1m high above the ground. In the first course, the laying starts with the thinner logs but the wood sizes become bigger as the kiln is built up with the larger pieces lying on top. During kiln building, an opening (fire box) of about 10cm x 10cm has to be provided on the leeward fire for firing the kiln.

Piling of Limestone: Well-sized limestone pieces are piled on top of the firewood layer. The limestone domes are 50-75 cm high. To prevent stones from sliding down, limestone pieces are held in position by pieces of iron sheets.

Firing: It is important to take into consideration the strength of the wind before igniting the firewood. Too much wind will aggravate the fire, finishing the firewood before the stones are ready. It is equally important to determine the condition of firewood, half-dried firewood is preferred for it burns slowly. Depending on the wind direction and speed, the desired temperature can be attained between six and twelve hours. If there is no wind at all, it can take up to 24 hours to burn the limestone, thus slowing down the process.

Slaking: Once the limestone pieces are fully burnt, the calcinated limestone is left to cool and later sprinkled with water for hydration. Iron buckets are used to pour water onto the heap of quick lime. One kiln load consumes up to 2,500 litres of water. After slaking, a spade is used to mix the product after which it is covered and left for one to two days to cool.

Screening: Two days after hydration, lime is sieved to remove unburnt stones.

Bagging and storage: After screening, the lime is bagged and the bags stitched and stored ready for marketing. Usually, 300 5 or 10kg bags can be packed manually by two persons within two hours using spades.

Investments
The Wachumico investment is as follows:
Office building (temporary)	TSh 50,000
Bank account (operating)	TSh 245,000.
Total	**TSh 295,000**
	($590)

Labour
There are no full-time employees working for the society. All jobs at the site are carried out on contract terms. Labour is based on specific tasks. For example:
(1) 10 people cut down a tree (mango, cashewnut, coconut) and size the firewood. The labour cost is TSh 10,500 ($ 21) for up to three days work.
(2) Transportation cost of firewood is TSh 9,500 ($ 19) per 7 tonnes
(3) Sizing of limestone is a five-hour job, estimated at TSh 500 per 2-tonnes ($1)
(4) Piling of stones on the firewood costs TSh 2,000 ($4)
(5) Setting up a clamp including piling of firewood cost TSh 3,000 ($6)
(6) Fetching water for hydration costs TSh 20 per 20-litre bucket.

Most Wachumico members have only primary school education; very few have secondary education. Their lime production knowledge was obtained through on-the-job training.

Production costs
The breakdown of production cost per batch is as follows (TSh):
Wages	9,000
Raw materials *(matumbawe)*	
per 2 tonnes	7,000 *or*
per 1-tonne	6,000
Energy (fuelwood)	10,000
Transport	15,500

Policies and regulations
Wachumico are licensed to produce raw materials, production and selling of lime. Tree felling does not require licensing.

Tanzania has a free economy with no price control regulations. Members of the co-operative society are self-employed and their wages are therefore, not statutorily regulated.

At national level, there are health and safety regulations (Factories Ordinance of Tanzania Cap.297). Some of the regulations are commonly known but not followed by this group. For instance, it is common knowledge that labourers should be supplied with glasses, gloves and helmets, but they are not. The reason given for not adhering to this regulation is affordability. Indeed most of the items identified are expensive, therefore cannot be afforded by individuals or the co-operative.

Waste
The amount of wate material produced is not known. There is about 2kg ash from the firewood used, but this ash mostly gets mixed in with the lime on hydration and is not normally separated. No measurements were made of oversize material discarded from screening the lime.

Markets
Agents of the society go to the city centre (Dar es Salaam) to look for market for the products. Usually a 10% fee of the total amount yielded from the sale is charged by the agents.

Due to the high demand for lime, marketing of the product is not a problem and all of it is always sold. The main consumers of the Wachumico's lime are building and hardware shops. A five-kg bag sells at TSh 150 while a 10kg bag sells at TSh 350 wholesale. Production, sales, and product demand records are not kept.

Main problems and bottlenecks
- The quality of the product is low and cannot satisfy different users.
- The producers lack skills and appropriate modern technology.
- The capital investment is too small to allow uninterrupted production, thus the level of production, productivity of labour and profits are generally very low. In this respect, production is done at subsistence level.
- Transport is very expensive, especially ferrying firewood logs to the production site. Boat owners cannot afford engines for boats such that, if there is no wind, boats take too long to reach the islands which have limestone deposits.
- Members/labourers face grave health hazards due to lack of protective devices.
- Lack of permanent land for office accommodation and production activity.
- In the absence of fixed assets and almost no floating assets, it is not easy to secure loans, because of lack of collateral.

Plans and prospects
- Wachumico aims to purchase their own 7-tonne truck to solve the transport problem.
- The society plans to apply for a permanent site from the Government.
- Members hope to send a few members for training who will in turn train the others in the co-operative.
- The society aims to secure loan facilities.

- The society is making efforts in applying to donor agencies to support them to inject some capital so as improve their technology and production generally.

Kigamboni Lime Works, Tanzania

Robert D. Reuben

Kigamboni limeworks is an integrated plant for production of lime. It is privately owned. The project is located in the Kigamboni area about 10km from Dar es Salaam city centre. The area is serviced with electricity, tap water and an all weather road.

The plant consists of a vertical (updraught) shaft with a capacity of 30 tonnes per day, a shaded concrete floor for hydration and a shed for screening, weighing and packaging. Other items include an office building, servant quarters and reserve tanks for water and fuel. The construction and installation of the factory was completed in January, 1993. Trial production started in February 1993 and lasted until August 1993. To date, full-scale production has not started. When the plant was visited, there was no production taking place. The information for this case study is based on the results obtained during the trial production period in 1993.

Raw materials

Limestone: Initially, the idea was to use limestone which is found in the neighbourhood. But later it was discovered that the local limestone:

- required high temperature for burning which meant high fuel consumption and thus higher investments
- gave low production efficiency, because it is porous.

For these reasons the owner opted to using limestone from Mwongozo about 10km away from the plant.

The quality of limestone, from both the neighbourhood and from Mwongozo is good with a very high calcium carbonate content.

Both samples slake vigorously and yield high calcium lime, suitable for industrial use in sugar, and leather processing as well as in building and civil engineering works.

Water: Tap water of a quality suitable for human consumption is used in production.

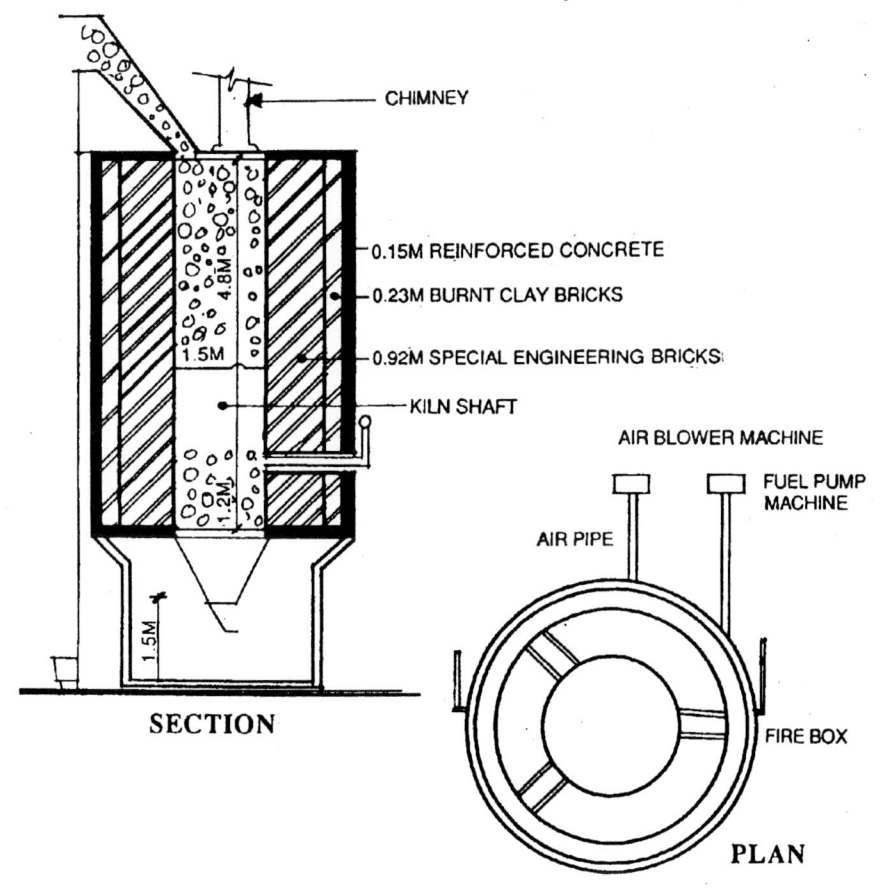

Figure 7: Vertical Shaft Kiln at Kigamboni Lime Works.
Note: Dimensions given were measured on site by the researcher and are not exact

Production technology and process

Extraction: Limestone is quarried from rock and dressed by using crowbars chisels and hammers. The plant owner is planning to use explosives (dynamite) in the future. Three people could produce greater or equal to 10 tonnes of limestones per day (10 hours). This would allow a reserve of 20 tonnes to be maintained at any particular time.

Transport: The owner has a 7-tonne truck which is used to transport lime stone from Mwongozo to the site. Energy needed for transport is 10 litres of diesel per trip. However, limestone from the neighbourhood could be carried using wheel barrows.

The lime kiln

The kiln is a vertical shaft type. It is designed to use industrial diesel oil.

The kiln is built on a firm foundation which is designed to carry the shaft and kiln contents. The foundation is a conical platform built on a concrete slab. The wall of the kiln is constructed in three layers, as shown in the diagram.

Fireboxes: At a height of about 4.8m from the top, three fireboxes are provided, equally spaced at the circumference. Size of each firebox is about 230 mm x 230 mm. In each firebox a pipe is provided which supplies fuel oil and air (oxygen) through the orifice at the inner end at pre-determined rates.

Air and fuel are supplied from different sources, but combine in a single supply when entering the fireboxes. The air is from the air blower, while fuel is pumped from the supply tank. The supply tanks receive the oil from a reserve tank. Control valves are used to stop and adjust the flow rates of air and fuel.

All equipment used was fabricated in Tanzania, and the design came from a kiln at Tanga. There is also provision for a chimney.

Production process

The production process and technology at the kiln involves the following stages:

Loading the kiln: The kiln is loaded from the feeding mouth which is situated at the top. The kiln capacity is 30 tonnes. The feeding is done using a power driven bucket hopper.

At the top the hopper discharges the limestone in a U-shaped thick mild steel sheet which is inclined down into the feeding mouth of the shaft. Due to gravitational force, the discharged limestone rolls gradually into the mouth, feeding the shaft to the required depth. Each charging and discharging of the hopper takes about five minutes. The capacity of the hopper is about 100kg. One person is involved in the operation. The stones fed from the top gradually progress down the kiln. The top layer is pre-heated by the exhaust gases and the air intake below is pre-heated by the cooling quicklime thus achieving efficient use of the available heat.

Calcination: The firing of the limestone in the kiln is done by using industrial diesel oil. During the process, the oil from the reserve tank is forced through the nozzles provided at the end of the service pipes into firing boxes by using an electric power-driven pump. With a good supply of air blown into the firing boxes to the nozzles, the fire is ignited to start gradual heating of the shaft to the highest temperature required. The maximum temperature reached in the kiln is believed to be above 1000°C.

The first discharge of the calcinated limestones is done after eight hours. The reason being it takes time for the temperature in the kiln to spread. Subsequent discharge are effected at four hours intervals. The amount discharged at each interval is 10 tonnes. With the continuous process an equivalent amount of stone is put in at the top of the kiln. Fuel consumption for calcination is 600 litres per 24 hours. The output per 24 hours is about 30 tonnes of limestone. Manpower required to run the kiln is five people per eight hours shift.

Unloading of kiln: The lime is discharged from the kiln by means of shovels from one discharge opening. The discharging mouth is mild steel sheet which is fabricated in a truncated inverted cone. It is about 1.5m above the floor. The quicklime is then carried by wheelbarrows to a concrete slab which is shaded ready for slaking.

Hydration: The discharged limestones are usually spread on a concrete floor under shade, for cooling. Every 12 hours, the quicklime is sprinkled with water, in order to yield lime. The job involves two people per shift of eight hours.

Screening and milling: The sieving is done using wire mesh sieves of aperture 150 microns. The lime is then milled in a machine to market grade fineness.

Bagging: The processed lime is packed into 25kg bags. A spring balance is used to monitor the weight and shovels are used to fill the bags. Stitching of the bags is done manually. Manpower required is six people.

Storage: The bags packed are transported everyday to Buguruni, about 15km from the site for storage.

Investment

Full investment on the Kigamboni Lime Works is valued at TSh 4,590,000 ($ 9,000).

Item	Amount(TSh)
Building	2,012,500
Machinery/equipment	2,437,500
Furniture and fixtures	100,000
Pre-operational expenses	50,000
Total	**4,590,000**

Forty per cent of his investment came from the National Bank of Commerce as a loan.
Note: $1=TSh520.

Labour

Full-time employees	Number
Managing Director	1
Production Manager	1
Machine Operator	1
Drivers	4
Watchman	1
Part-time employees	
Quarrying	6
Spreading of discharged limestone	2
Hopper bucket feeding	2
Sieving and packaging	6
Stitching	1

The full-time employees are of secondary education with some technical training. The part-time employees possess primary school education and have gained experience through on-job-training. Wages for full-time employees is above the national minimum wage which is TSh 10,000 ($2) a month. However, the salaries and actual figures were not revealed due to unknown reasons. Part-time workers are paid weekly on a piece-work basis.

Production costs

The production costs per year were given in a summary as follows (TSh):

Item	Cost
Labour	125,500,000
Fuel (diesel)	12,000,000
Packing bags	78,000,000
Electricity	12,000
Transport	147,032
Total	**147,032,000**

Waste

The amount of waste from lime quarrying has not been established. Notable waste is however, obtained during sieving. The figure of waste is around 10-15%.

Marketing

Theoretically, this is the only large production unit for lime in the Dar es Salaam region. The lime is also of high quality. Bearing in mind that Dar es Salaam is the largest consumer of lime in the country and a ready market.

Capacity utilization

The plant has the capacity of 3,000 tonnes per annum per single shift (eight hours). However, capacity utilization of the factory is zero since in the past one year, no production of lime has been recorded. This is a sad state of affairs considering that in our estimation, the annual national demand of lime stands above 100,000 tonnes.

It is unfortunate that keeping good records is not a high priority, for we could not secure any sales figures showing the marketing of lime produced during the time when the kiln was on trial production.

Oil-fired kiln under construction at Kigamboni Lime Works

Policies and regulations

Generally, there are national rules and regulation to be followed by factories or small-scale industries. The case study revealed that most of the rules are not known to the owners, therefore not adhered to. Some of the rules known by the owner of the plant are:

- Insurance rules: the owner is expected to insure his plant and employees
- All industrial businesses are required to acquire industrial license for industrial production to take place and business license to run the business and licence for lime stone quarrying.
- The factory ordinance Cap.297, provides that: workers must be protected from inhaling dust; they ought to wear gloves, gum boots and helmets. Fire extinguishers are also necessary within factory premises.

Some factory owners supply milk to their workers. Mr. Asenga, the plant owner, is also planning to provide such facility when production resumes.

Pricing regulations

The economic system is a free market. There are no regulations controlling pricing of commodities. Producers can fix prices of their products in accordance with rules of supply and demand (market forces).

Wage regulations

Mr Asenga's kiln is privately owned, as such he can decide on the rate of salaries to pay his employees as long as the salaries are greater or equal to the national minimum wage.

Health regulations

In accordance with the National factories regulation ordinance, Cap.297, the rules must be complied with. The plant is not operating currently and it is not known if relevant health regulations were complied with in the past.

Environmental regulations

Currently Tanzania does not maintain any law specifically dealing with environmental matters. However, it is anticipated that a comprehensive environmental protection bill would be passed as law before the end of 1994. Presently the factory ordinance has a number of provisions which provides for the environment such as the requirement to take precautions with respect to explosive or inflammable dust, gas, vapour or substance and removal of dust and fumes. However, these are quite unreasonable.

Problems and bottlenecks

For a period of one year, the plant has not been working. Rising costs of fuel and electricity make it very expensive to run the project. There are also insecurity problems and within three months from March to May 1993, an oil pump, milling machine motor and discharger motor were all stolen from the plant.

It is sometimes difficult to get lining bricks from the Wazo Hill Cement factory of Tanzania when needed to replace the lining in the kiln. This makes it necessary to import the lining.

When the project started in 1993 there was an electricity rationing exercise in the country. It was not possible to run the machines as planned, thereby negatively affecting production.

Plans and prospects

- It is planned that the plant produces additional products, i.e. a special fine lime for manufacturing tooth paste.
- Due to anticipated cost increases of oil and electricity, the management is planning to use coal instead.
- It is also planned to extend the height of the kiln shaft to 7.5m (25ft), to improve burning efficiency.
- The management is willing to have a joint venture with any partner who is interested and who is a lime production expert.
- The owner is thinking of having a multi-purpose crusher (jaw) for stones.

All of these plans may be wishful thinking unless concerted efforts are made to address the current problems of the kiln/project.

Amboni, Tanga District, Tanzania

Robert D. Reuben

There is a great demand for lime in Tanzania, which finds many uses in industry and in agriculture. Lime produced by Tanga Lime and Super Amboni Lime factories (the major lime producers in Tanzania) cannot meet the demand. Tanzania and especially the Tanga District is endowed with rich deposits of high quality limestone. To date not much of the deposits have been exploited.

For this reason an area of high quality limestone totalling 30,000 m has been leased. Geological surveys conducted on this land show that the carbonate content of the limestone is 93-98%. A simple lime producing unit has been started to minimize the gap between demand and supply of lime in the country.

Raw materials
Geological surveys show that there are deep deposits of limestone rock of excellent quality in the leased area. Drilling explorations revealed that there are limestone deposit up to a depth of 250 metres.

Quarrying
Limestone is obtained from open-pit quarries. The advantage in the leased area is that rich high calcium rocks are only a few centimetres below ground level. Only when the open-pit quarries are exhausted would underground mining be done. In order to obtain crushed limestone for the kiln, the following series of operations are necessary:

- drilling holes for explosives
- blasting the limestone loose by use of explosives
- crushing and sorting the stone to a uniform size of 6-8cm
- loading the stone into lorries
- transporting the stone to the burning site

Batch production at Amboni, near Tanga

Production technology and processes
The blasted limestone rock is crushed by hand hammer to the required uniform size then loaded on a heap of firewood which is set on fire. A burning temperature of between 900°C and 1,100°C is maintained for two days.

Slaking is effected in layers of burnt lime some 10-12.5 cm inches thick by sprinkling it with water from a watering can. Mixing the material is done with a shovel while adding water. The material is left for a day so that the slaking may be completed. After slaking, the material is sieved to separate overburnt and underburnt material from the slaked lime $Ca(OH)_2$. The fine material is then packed in paper bags ready for sale.

Investment
Capital was needed to cover (TSh):

Acquisition of leased areas	16,000
Survey fees	12,000
Prospecting Licence	5,000
Registration fees	6,000
Claim fees	6,000
Business licence	12,000
Equipment and tools	280,000
Construction of shed, site clearing, etc.	65,000
1st trial operation	170,000
Total	**572,000**
	(US$1,140)

The investment was obtained through own savings. This also included the start-up operational costs of about TSh 183,000 ($360).

Labour
At least 4 people are required for the preparation of fire logs on a heap in the shape of a circle. One of the four people is an artisan who is responsible and skilled in heap-burning and the other production processes. The other three merely assist the artisan. During the slaking process several women are employed to ferry water due to lack of a water supply at the site. About 4,000 litres of water are required. Here 4 to 6 women are engaged. Two women carry out the sieving using one sieve. Production is 50 to 70 bags of 25 kg per day.

Skills and qualifications
There are no specific qualifications required for the heap burning. Anyone who is attached to this work within a reasonable time can do the work. The work is seasonal. The dry season and hot weather are the best times for heap burning.

Payment system

Payment is normally fixed on a task basis. The artisan is paid TSh 20,000 ($40) per heap burned. The other three attached labourers are paid TSh 300 ($ 0.60) per day.

The sievers are paid TSh50 ($0.10) per bag of 25 kg. Women who collect water are paid TSh 25 per bucket of water. Payments are made after completion of the task (one heap burning). Usually it takes seven days to produce 10 tonnes out of one heap burning. The workers are also paid social and other fringe benefits to motivate them. Although there is a high manpower potential in Tanga Municipality, the local people have a very low attitude towards blue collar work, so are reluctant to take up these type of jobs.

Production cost per week (10 tonnes)

Salaries & wages

Administration	7,000	
One Artisan	20,000	
3 Labourers @ 300 per day	6,300	
Sievers 50 x 400	20,000	
Collecting water	5,000	
2 watchmen @ 350 per day	4,900	
Sub-Total	63,200	
5% Social Benefits	3,160	
Total	**66,360**	**($133)**
Material costs		
Fire Logs	70,000.00	($140)
Explosives, quarrying, crushing & transport	14,000.00	($28)
Packing materials 80x400	32,000.00	($64)
Total	**116,000.00**	**($232)**

Waste

Waste in terms of limestone is 25%. The waste which is underburnt and overburnt limestone is crushed by hand to small sizes and sold as aggregates. A smaller amount of wastes is carried away by wind during sieving. At the same time about 15% is waste (course material) which does not pass through the sieve to the required fineness.

Markets

All lime produced is easily sold by the wholesalers. Wholesale prices are TSh 500 ($1) per bag of 25 kg, and TSh 20,000 ($40)per tonne.

There are many simple lime production units in Tanzania, but they do not meet the national demand. The price of imported lime is substantially higher. Competition in this industry is hence very minimal. Most small producers do not produce a high quality lime due to either the quality of the limestone or poor production technology. Site purchase is encouraged; otherwise the product is distributed to users at an agreed fee.

Policies and regulations

Licences

A dealers licence from the Department of Minerals is necessary before one can operate a lime production unit. The licence is renewable annually. Apart from that, a normal business licence is also necessary. In order to obtain the above licence the income tax department has to issue a tax clearance certificate.

Pricing regulations

There are no specific pricing regulations on lime products. However, a standard bag of lime should weigh 25 kgs. A 25 kg bag costs TSh 500 ex works.

Wages and health regulations

Wages are paid in accordance with government fixed statutory wages to casual labourers and temporary employees. For health purpose sievers are provide with one litre of milk each day. They are also provided with gloves to avoid occupational diseases.

Fuelwood supply

There are specific government policies which govern cutting of trees which in the end might lead to non availability of fire logs from the forests. Hence the urgent need for an alternative source of energy for production.

Main bottlenecks and problems

These can be listed as follows:

- supply of fire logs
- availability of packing material
- shortage of explosives
- loans and overdraft from financial institutions. Banks require real estate as security for loans/overdrafts. The interest rates are very high (about 40%). The process of obtaining loans/overdrafts is also very lengthy.
- transport

Plans and prospects

The level of production outlined above is intended only to be a temporary measure because market potential indicates that a continuously operated 50-to-75-tonne-per-day plant would be viable. This plant would be relatively highly mechanized and an air spearator will be used to separate out the lime into various-sized fracions. The plant would be able top supply various industrial users of lime as well as the building market.

Village Lime Kilns: The SIDO experience in Tanzania

E.G.S. Ikomba

The Small Industries Development Organization (SIDO) was established by Act of Parliament No. 28 of 1973, following the Party, Policy guidelines on small-scale industries issued in April, 1973. The organization started working in the first half of 1974.

The act stipulates the following functions for SIDO:

- to promote the development of small industries in Tanzania
- to plan and coordinate the activities of small industry enterprises in Tanzania
- to carry out market research in goods manufactured by small industries in Tanzania
- to advise the Government on all matters relating to the development of small industries in Tanzania
- to provide technical assistance to persons engaged in small industries
- to undertake or assist any institution or person in the undertaking of technological research and to encourage, and promote technological advancement in Tanzania, etc.

At National level SIDO is responsible for planning, coordinating, promoting and offering technical economic and management services to both existing and prospective small industries through its network of 20 Regional extension offices. SIDO also extends consultancy services in the field of technical, economic, management, marketing, production and quality control, know how and technology transfer, and conducts regular multiskills training programmes, implant studies and techno-economic surveys. It has technical personnel, including specialists in specific fields of operations.

SIDO has been involved in the promotion of artisanal industries with special focus on the rural environment; some of the trades are offered at its training cum production centres. So far the centres have trained over 4,000 artisans in a number of artisanal trades including:

- village lime burning
- lime pozzolana production
- burnt bricks and tile manufacturing
- cement soil blocks production, etc.

Village lime kilns

The Village Kiln Programme was introduced by SIDO in the second phase (1976 to 1978) of its development work. The objective of this programme was to establish small-scale industries at a village level which use locally available raw materials, as well as to create more employment in rural areas and to retain youths in the rural areas after the completion of their primary education.

An advisory mission by ITDG, in August 1974 suggested that the Indian experience with village lime technology should be transferred to Tanzania. Subsequently, with Indian help a training programme on lime burning and kiln construction was conducted in 1976 for 25 participants. The aim was that after the completion of three months training, they would go back to their respective regions and train the villagers on the technology acquired.

Afterwards, SIDO helped to establish 30 kilns across the country. Most of these were financed/donated by SIDO or funded from the regional development funds and about five projects were assisted by the British Council, CUSO and religious institutions.

Raw materials

Almost all regions of Tanzania have got limestone deposits with varying degrees of quality. Villages get their limestone from their own quarries, without payment. They also use firewood cut from nearby forests which is mostly free of charge.

Production technology and process

The choice of technology which was used in lime burning is simple with the aim of minimizing the need of skilled manpower to operate and maintain the plant. SIDO advised on the available simple, low output technology (kilns and methods of slaking) which had been sought from the Khadi and Village Industries commission (KVIC) in India. The first kiln to be set up was a vertical shaft kiln with a capacity of five tonnes per day. The basic design is such that, if necessary it can easily be modified to suit the different output requirements, also according to the nature of limestone and fuel used. It can be operated on a batch or continuous basis with coal, charcoal, or firewood as fuel. This kind of kiln is easier to construct and locally available materials can be used. The kiln has to be fed in alternate layers with limestone and firewood. Simultaneously the calcined quicklime lumps are drawn from the base of the kiln.

Slaked lime can be used in building construction, as a putty or in powdered form. The simplest

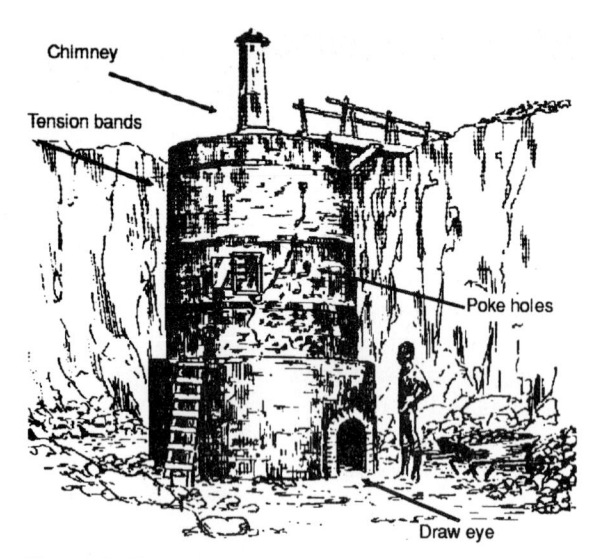

Figure 8: The KVIC-type kiln, a number of which were constructed in Tanzania

method of slaking is by spraying water over the quicklime lumps on a slaking floor; the slaked lime has to be screened by a trommel screen then bagged ready for sale.

The lime kiln is constructed from burnt clay bricks and has to be plastered internally. The loading of the kiln is done by first placing coconut shell as the ignition material over the grating bars, then fuel and stone in alternate layers. The kiln is not, however, filled completely to the top to ensure good draught. A gap of two feet from the top is left unfilled inside the kiln for quick escape of gases.

To start the kiln coconut shells are piled just below the grating and ignited. After ten minutes the shells ignite the wood in the kiln. All the underburnt lumps retained on the screen are charged back in the kiln.

This method uses locally manufactured tools and equipment. Supervision of kiln construction and training for kiln operators and installation work was conducted under the technical guidance of SIDO.

Investment

Most of these kilns were financed by SIDO or assisted by other institutions. The production capacities of these kilns is 3-20 tonnes/day; and construction costs range between TSh 1.2m to TSh 2.0m ($2,400-4,000), depending on the location. These plants are owned by the village communities.

Labour

Every village has its own programme of lime burning; some units work once a week while other units work twice a week and they work on a self help basis, under Ujamaa policy. No salaries are paid after the job is done. The money received from sales belongs to the village. Most of the workers are standard seven leavers, or mostly illiterate adults.

Current situation

As stated earlier out of 30 lime kilns set up in Tanzania under the technical guidance of SIDO, only 10 lime kilns are still working. These are: Hombolo and Mvumi Mission (Dodoma); Maweni Prison and Kiomoni in Tanga; Makere Kasudeco and Simbo (Kigoma); Vitono; Kiwere and Ikengeza (Iringa).

These kilns work as production wings of the villages. The 'Ikengeza Lime Unit', for example, is located in Iringa region along the Iringa-Dodoma highway, about 45 km from Iringa municipality. The project belongs to Ikengeza village. It employs 10 people and they produce between 20-25 tonnes per week. Lime produced is being sold at TSh 250 ($0.5) to 600 ($1.2) per bag of 25 kg or at TSh 18,000 ($36) tonne. The workers are well paid and paid on a monthly basis; at the same time they retain 20% of the sales income. Because the workers are paid a salary rather than working on a self-help basis, this is considered why this particular plant has been more successful than some of the others.

The project has assisted the village with the construction of a health centre and a nursery by providing them with building materials.

Policies and regulations

Some villages have no licence for raw materials and wood cutting from the forest, though they have been burning lime for quite a long time. There are no health and safety regulations being implemented.

Main bottlenecks and problems

Production in these projects is affected because often there is no payment of salaries or wages made for the work done, thus discouraging villagers to be more effective and produce more. Lime producers need to be employed by the village management and paid regularly; this will enable them to produce more and have higher sales/returns, as in the case of Ikengeza. In addition, accounts are not properly maintained. Projects are affected by illiteracy, lack of information and general poverty.

Plans and Prospects

- All lime kilns which are in bad technical shape to be repaired/rehabilitated.
- All lime projects needs at least someone with knowledge of bookkeeping.
- In order to produce more lime the system of working on self help basis should be stopped and instead workers have to be paid salaries or wages on regular basis.

To achieve these, there is a need for the involvement of Government and development organizations in small-scale lime production. Lime production could be considered more viable if producers could also make lime-pozzolana cement using burnt clay, rice husk ash or volcanic ash pozzolana.

Kassala area (Al Gira), Sudan

Mohammed Elamin Gasim

Kassala is one of the major lime production areas in Sudan. Traditional lime burning around Kassala town is said to go back over 200 years. There are about 60 producers around Kassala in five areas of production as follows:

Area	Location[1]	Number of producers	Number of kilns
Al Gira	25 km SE	27	55
Kabudye	28 km SE	8	19
Diman	20 km E	3	7
Amanit	32 km NE	10	12
Shalalobe	35 km NE	12	22
TOTAL		60	115

1. Relative to Kassala township

The main area of lime production in Kassala State is Al Gira. Al Gira has a population of around 2500 (80% Sudanese and 20% Ethiopian and Eritrean) of which about 400 workers depend on lime burning for their main income. During the rainy season workers and their families are supported by small land holdings on which they cultivate some of their annual nutritional requirements.

The kilns are located alongside the eastern bank of the seasonal river Gash. The Al Gira area provides skilled kiln operators to other lime burning areas of Sudan, as they are generally considered to be the most experienced. But currently, The Al Gira area suffers from a shortage of skilled labour because, as the workers grow older, their sons are not taking up the profession, preferring to go into towns where they find better rewards and good working conditions.

In Al Gira, there are about 54 kilns belonging to owners from outside the area. They employ casual labourers to operate the kilns. Labour is organized around skills with around 14 people specializing in building and loading the kilns (including the owners) and another 14 in firing kilns. Over 300 people are employed in various manual jobs and most work is done on a piece work basis.

Production of lime in Al Gira is concentrated in the period from November to May with only two or three producers operating throughout the year. The average production per year is five batches per kiln; each batch averages approximately 25 tonnes of product (quicklime).

An example of lime production

The kiln taken as case study is located in Al Gira. It is owned by a retired army officer. He is operating the kiln from his own financial resources without any assistance from financial institutions. The owner has the capacity to perform up to five firings per year.

Licences

To obtain a licence, producers need to define their area of operation and ensure it is not owned by somebody else. They must then get approval from local authorities including:
- The State Ministry of Agriculture (to check that it does not affect agricultural land)
- Forestry Department (to ensure lime production is outside forest reserves and far from fire lines)
- Municipal Council (to issue the licence)
- Health Department (to ensure lime burning does not affect residential areas nor pollute the environment)
- The committee in the central department of geology and mining.

The licence fees are Ls 25,000 per year in addition to Ls 6,000 local fees (US$60 + $15).

Raw materials

There is no information on the extent of limestone reserves. Limestone sedimentary deposits are quarried from two locations in Al Gira.

- The bed of the seasonal river Gash, where quarrying is done during the dry season only. Deposits are shallow and occur between layers of sand. Though it is easier to quarry, the quarried limestone is expensive due to transportation and labour costs.
- Adjacent to the kilns, where quarrying is done throughout the year. The quarries have around two to three metres of overburden which must be removed prior to winning the stone. The stone is much cheaper to quarry then that from the first location due to low transport costs.

Unproven observations indicate that there are sufficient reserves of good quality limestone in the area for many years of artisanal production. Al Gira limestone is a course crystalline white marble. Samples of limestone from the kiln under study have been analysed with the results shown below.

In this particular case, the limestone is owned by the kiln owner. Other producers lack limestone deposits on their lands so they buy quarried limestone from the river Gash, at Ls 5,500 per tonne ($13), including transportation to the kiln site.

Constituent	Sample 1 %	Sample 2 %
SiO$_2$	13.89	12.84
CaO	38.50	57.77
MgO	9.62	10.38
Al$_2$O$_3$	0.78	1.05
Fe$_2$O$_3$	0.44	1.02
LOI	6.06	15.42
Total	**99.29**	**98.48**

Production technology
The lime kiln
The lime kiln is cylindrical in shape, typically 4.2 metres high and 4.2 metres in diameter. The walling is 0.35 metres across and built of fired bricks. A layer of mud, bricks and limestone is placed all round the outside to reduce heat loss during firing. The top of the kiln is open and there is a 1.5-metre high door on one side for access. The bottom of the kiln consists of a 1 metre deep fire box below ground level, lined with fired bricks.

Kiln operation
The kiln operates on a 27-day cycle according to the following sequences:

Limestone quarrying and dressing: The limestone is quarried utilizing simple hand-tools (chisels, hammers, crowbars). Sometimes workers heat the deposits utilizing used rubber tyres. One person could quarry 1.5 tonnes of stone per day. The whole kiln batch requires 42 tonnes of limestone (two labourers take 14 days to quarry this quantity). About 8 tonnes of this limestone is then broken further with hammers to large lumps of up to 0.4 metre long to form a dome. To build the dome a gap of 1.8 metre from ground level is left at the door opening (a steel lintel is used to span the door and support the dome). The remaining 34 tonnes of stone is broken to kiln feed size of minus 50 millimetre in diameter. (It takes 3 days to build the dome and to load the kiln with limestone.)

Kiln firing: The kiln firing commences when the kiln is loaded and the fuelwood is on site. The first operation is to fill the fire box and it is refilled 3 to 4 times a day for 4 days whilst firing. Firing requires 3 labourers.

Cooling and unloading the kiln: After firing, the door is bricked up and the kiln left to cool for 3 days. The 2 labourers require 3 days for unloading the quicklime.

Packing quicklime: Used sugar sacks of about 10kg capacity are used for packing quicklime.

Fuel
The main fuel used is indigenous acacia and dome (palm) trees, brought over illegally from Eritrea, but these are becoming difficult and expensive to obtain. Currently, the main source of fuelwood in the area is mesquite which is in plentiful supply. The mesquite tree is exotic to Sudan being originally introduced as a means of controlling soil erosion and advancing desertification. It has been so successful that it is now considered as a pest and is invading good agricultural land.

In this case study, due to difficulties in obtaining acacia or mesquite, the owner of the kiln purchased old non-productive fruit trees (grapefruit), one tonne of which costs around Ls 8,500 ($20), including felling cost and transportation.

Inputs and outputs
A detailed physical measurement of a batch was done; this measurement indicates that:
- the quantity of limestone used for this batch was 42 tonnes
- fuelwood used for this batch was 18 tonnes
- actual output of quicklime was 25 tonnes
- the waste in production was 12 tonnes.

Production costs
The capital cost of kiln building is:

Materials:	Ls
12,000 burnt bricks	144,000
4 steel lintels	120000
Limestone	15,000
Wages:	
Excavation	26,000
Building	115,000
Total	420,000
	($1,000)

The variable cost of one batch is:

	Ls	%
Limestone quarrying	70,000	23.37
Limestone dressing & loading	275,000	9.18
Dome building	7,000	2.34
Fuelwood (18 tonnes)	153,000	51.09
Firing	30,000	10.02
Unloading	12,000	4.00
TOTAL	**299,500**	**100%**
	($720)	

Marketing
Producers of lime concentrate on quicklime production because local consumers prefer quicklime, thinking that an indicator of good quality is its reaction with water. The prices of the quicklime have been increasing dramatically during the past years due to the high costs of fuelwood, transportation, wages and government fees.

The current retail price of one tonne of quicklime is Ls 55,000 ($130) while the price of Portland cement is over Ls 60,000 ($145). (Note: After slaking, a tonne of quicklime should produce at least 1.3 tonnes of lime hydrate, so the latter's cost is less than two thirds of the cost of cement.) The producers sell their quicklime on site without packaging at a price of Ls 28,000 per tonne ($66).

The production of quicklime in Kassala is estimated to be around 14,375 tonnes per year with Al Gira producing approximately 47% (6,750 tonnes).

Main bottlenecks and problems
Lime production in Al Gira area suffers from several different problems including:
- high fuelwood consumption rate
- exposure of workers to high temperatures and excessive lime dust during kiln firing, unloading and packaging
- poor management and a shortage of skilled labour
- lack of financial and technical assistance

Future plans
Future plans include:
- improved access to loans for expansion
- kiln design development to improve efficiency and productivity.

The lime industry in Malawi
Otto Ruskulis

Lime processing in Malawi is a traditional activity, largely concentrated around the Balaka area. Balaka is a medium-sized town on the main Blantyre to Lilongwe road. Over the past 10 years, largely through the efforts of Intermediate Technology, improved lime production technologies based on continuously operated vertical shaft kilns have been introduced into the area. The majority of lime producers in Malawi are members of a trade association, the Lime Makers Association of Malawi. Most lime producers are likely to derive between 20 and 60% of their incomes from lime. Employees of traditional lime producers would make between about 30 and 50% of their income from lime production, or around 90% of their cash income, excluding the value of crops which they grow themselves. The remainder of the incomes of lime producers or employees would be made up on activities such as agriculture, fishing and casual work such as road repairs. Some of the kiln owners also have one or more trucks and earn an income from transport.

In the early 1980s, the Lime Makers Association approached the Intermediate Technology Development Group for assistance with developing improved lime production technologies. A number of problems had been identified with traditional techniques including poor energy efficiency, deforestation of indigenous forests and high volumes of waste produced which could not be turned into useable lime.

Raw materials
The areas where lime has been produced in Malawi include the Balaka area and a small area around Uliwa in the north, close to Lake Malawi. Intermediate Technology provided technical assistance to set up a lime production plant at Uliwa in the early 1980s based on a vertical shaft kiln. The stone was of high quality, it passed through the kiln without breaking up, slaked well and did not need to be ground after slaking. However, the plant appears to have ceased production in 1993. The reason for this is probably lack of a market. The north of Malawi is relatively remote, mountainous and sparsely populated. The plant is still, however, largely in place and production could be re-started if an imaginative way could be found to market the product such as persuading local builders to use lime in building.

The majority of lime deposits in the Balaka area are located at the Chenkumbi Hills, a range of small limestone hills some 10km south of Balaka. The measured reserves of the main limestone group stand at 3.7 million tonnes. The reserves are a coarse-grained calcitic marble of a sufficiently high chemical purity for production of chemical grade lime. Below is a chemical analysis of the limestone:

Calcium Oxide (CaO)	52.06%
Magnesium Oxide (MgO)	2.23%
Inerts and traces	2.69%
Loss on ignition	43.02%
CaO content of limestone	1.40%
Available lime content	69.70%

It has been found that the limestone at Chenkumbi has a tendency to decrepitate (break up) as it passes through a lime kiln. This has sometimes necessitated the use of a forced air blower on the vertical shaft kiln.

Quarrying is by manual methods using picks, shovels and hammers to break up the stone. Chenkumbi Limeworks hopes shortly to start to use explosives to bring down the stone.

Production
It is estimated that there are some 40 lime producers using traditional techniques in the Chenkumbi area and probably around 40 elsewhere. There is currently one lime plant in operation based on the improved technology. In Malawi between 3,000 and 3,500 tonnes of lime are produced by traditional methods, and just over 1,000 at the improved plant.

Traditional lime production
A diagram of the kiln traditionally used for lime production in Malawi is shown in Figure 9. This is a rectangular box kiln of about 2.5m^2 section and about 3m in height. Two trenches run through the bottom of the kiln and these are used for feeding in logs during firing.

Before firing the kiln an arch of limestone boulders is built over the two trenches. On top of these are put five alternating layers of fuel and limestone. Timber from local indigenous forests is used as fuel.

After igniting, the kiln is allowed to burn for 48 hours. The fuel is supplemented by periodic feeding of logs into the trenches. The top of the kiln is then sealed and the kiln left to burn out and cool. This takes about another eight days. Quicklime is removed from the kiln only when the need arises or when a sale for a quantity of lime is secured. This inevitably results in some air slaking.

Slaking is carried out by pouring water over the quicklime and turning with a spade, and the final product is usually sieved to remove the larger lumps. Most producers also pass the final product through a hammer mill and there are a small number of mill operators in the area. However, because the mills used are intended to grind maize rather than minerals and often contain mild steel beaters breakdowns are common and the grinding is not efficient.

Lime is put into sacks manually. Often second-hand fertilizer or relief food sacks are used because paper sacks are too expensive.

Typically around 70 tonnes of limestone and 50 tonnes of wood fuel are charged into the kiln. Most producers would fire the kiln five or six times a year, although this can be less if there is a reduced demand for the product.

The product has an available lime content of 40%, or less, which is low. With efficient production techniques a figure in excess of 60% could be expected.

Improved lime production: the UML plant
During the first phase of its project to provide technical assistance to improve lime production in Malawi, Intermediate Technology first built a small plant in the Chenkumbi Hills based on a small continuously operated vertical shaft kiln. This was never intended as a commercial plant, but rather as a pilot plant to test out the properties of the stone and the best way of burning them. Through these trials it was established that:

- the available lime content of the finished product could be raised to about 45%
- plantation softwood could be used as fuel instead of hardwood from the forest
- waste was reduced
- because of decrepitation (breaking up) of the stone in the kiln it was found necessary to use a fan to force air through the kiln to improve efficiency
- with the forced air system in operation the continuously operated kiln was about three times as fuel efficient as the traditional batch kiln.

At around the same time as the production trials were taking place a project was formulated to set up a commercial scale lime production plant. One of the main markets for this plant was going to be the sugar industry in Malawi which at the time was being supplied with lime from Zambia and South Africa. This project was to be managed by READY, a Malawian organization, with funding from USAID and technical assistance from Intermediate Technology. A private company, UML, was set up to run the plant as a commercial enterprise.

The kiln's internal dimensions were a height of 6.6 metres and diameter of 1.1m, tapering to 0.8m at the top. It had a number of notable features:
Refractory and insulation bricks: Locally produced refractory bricks for the kiln were developed with the help of a pottery workshop. The manager of the workshop was able to tell by experience and after some trials which of the local clays made the best refractory and insulation bricks.

Forced draught: for improved burning efficiency.
A steel shell: for structural stability and to reduce cracking within the kiln.
Loading of stone and fuel: at the top by means of buckets drawn up by a windlass fixed to a davit arm.
Discharging of stone: at the bottom from four openings using shovels.
Inspection holes: Four inspection holes were put in vertically along the side of the kiln, for visually assessing burning conditions and, if necessary, unjamming blockages.
Chimney: A three-metre-high metal chimney was put on top of the kiln for improved airflow.

Apart from the kiln, a range of other processing equipment was also developed for the plant:
A mechanical batch hydrator: This was not altogether a success. The hydration reaction was very violent and dangerous to anyone standing near to the hydrator. A continuous mechanical hydrator would have been a better option.
An air classifier: This was specifically developed for the plant. It enabled fine particles to be removed continuously from the feed to the hammer mill so improving grinding efficiency.
Dust filtration: Prototype dust filtration equipment was developed for the plant.
Bagging: A simple mechanical bagging unit was developed for the plant.

The total cost of the plant including land, buildings and working capital but excluding research and development costs was about $120,000. The kiln cost was about $18,000.

The plant was capable of producing up to five tonnes of lime per day. Wastage was very low and, at 36%, efficiency of the kiln was four times that of the traditional kiln. The main difficulty with the plant was the processing equipment, some of which was innovative and developed specifically. In particular the air classifier and hammer mill closed circuit required considerable modification before it worked satisfactorily. The modifications caused many months delay in starting production and a cost overrun for the project.

After about two years of production UML, the company which ran the plant, went bankrupt and the plant has not worked since 1993. There were a number of reasons for this, but one of the main ones must have been that the company was not able to capture a significant share of the sugar industry market for lime in Malawi. Presently (December 1994), a buyer is still being sought for the plant who could take it on as a going concern.

Improved lime production: Chenkumbi Limeworks
Chenkumbi Limeworks was established with considerably different objectives to the UML plant. The long-term aim was that the company would be owned by the traditional lime producers in the Chenkumbi Hills area who could use the profits from the plant to improve their own lime

production facilities. The plant would also provide other facilities, such as a grinding mill, which lime producers from the area could use on a shared basis.

For these reasons, and also because the main market for the product was to be the building rather than the sugar industry, whereby a very pure lime was not required, a considerably simpler and less costly design for the plant was adopted. The kiln is similar to the UML kiln although slightly taller and thinner. It also has a refractory lining and a layer of insulation bricks. However, the outer steel shell is replaced by a layer of building bricks. To reduce the likelihood of cracking a series of tensioned steel bands have been placed around the kiln. The loading mechanism has also been simplified. The kiln has operated more or less continuously since August 1993.

Although provision has been made for a forced air blower system, this has been found not to be necessary. An accurate assessment of kiln efficiency or waste produced has not yet been carried out but the indications are that these values would be similar to those of the UML kiln. The fuel mainly used is charcoal, which is now the cheapest option.

Hydration is done by sprinkling water on the quicklime on a concrete surface in one of the sheds. The lime is then ground in a hammer mill and blown through by a fan into a storage shed. After the grinding has stopped and the dust has settled the shed door can be opened and the lime removed for bagging. Bagging is done manually with shovels, and the weight of the bags is checked on a set of scales.

One further point to mention is that the plant has no mains electricity supply but it does have its own generator. Also, until very recently, it did not have its own water supply, and water had to be brought manually from a few kilometres away.

The total cost of the plant (including the electricity generator) is likely to be around $50,000. Production of the plant is now up to five tonnes per day and the order book looks healthy. In the first year of operation it made a small profit. To meet the demand for lime the company has sometimes to buy unground lime from some of the traditional lime producers which it then grinds and sells. The plant manager and a marketing expert, who has been working temporarily at the plant, have developed a strategy for getting the product into the main markets of the large towns of Blantyre and Lilongwe, which are some distance away from Balaka, using the existing transport networks. Most of the product is used in building.

Despite the commercial and technical success of the plant the take up of shares in it by small-scale lime producers in the Balaka area has been somewhat disappointing and Intermediate Technology still remains the majority shareholder.

Markets

The main market for lime produced in Malawi is building, largely for use in decorative and protective renders on walls. The sugar industry is the other main user of lime in Malawi but it has not been convinced that lime produced in Malawi is of sufficiently good quality or consistency to meet its needs, and imports from Zambia and South Africa. There is also a small market for lime for chemicals and paint making for which a high purity lime is required and which is met through imports.

Bottlenecks and problems

These can be listed as follows:

- depletion of hardwood forests, the only realistic fuel option for production by traditional methods, leading to growing fuel prices and scarcities.
- low fuel efficiency (only about 9% taking account of the waste produced) of the traditional kilns.
- poor quality of the product using traditional methods.
- good quality limestone reserves in the north are far from the markets for lime.
- traditional lime producers lack the skills and the means to effectively market the product.
- the Malawi sugar industry is still to be convinced that locally produced lime using the improved methods will consistently be of sufficiently high quality.
- few of the traditional lime producers have the means to acquire shares in Chenkumbi Limeworks.
- most of the traditional lime producers are unlikely to be able to acquire the capital to build a plant such as that at Chenkumbi Limeworks and they might not want to be full-time lime producers. For them low cost batch kilns which are more efficient than the traditional kilns and which can burn a greater range of fuels would be a more realistic option.
- restarting the UML plant as a going concern.
- rice is grown in Malawi, but not in sufficient

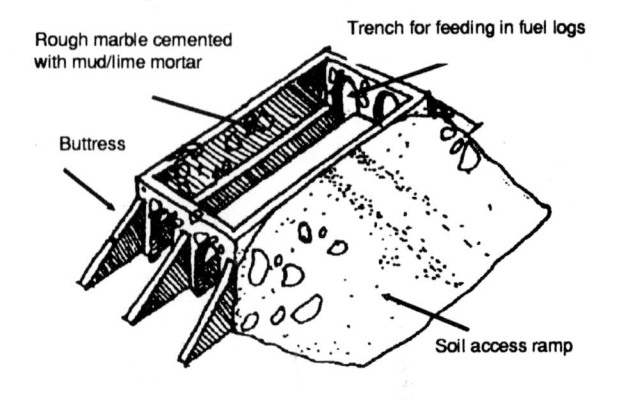

Figure 1: A traditional lime kiln in Malawi

The view over Chenkumbi Limeworks in Malawi

The continuous lime kiln at Uliwa in Malawi

quantities to be able to make use of the husks in making lime-rice husk ash cement to help relieve the shortage of cement in Malawi.

Plans and prospects
- The official receiver of the UML plant will look for opportunities to enable the plant to restart production on a commercial basis.
- More research is needed on improved low cost batch kilns capable of burning a variety of fuels to enable small-scale intermittent producers to continue in production as fuelwood reserves in the indigenous forests become depleted.
- A research project on optimizing fuel efficiency and reducing waste is to be carried out at Chenkumbi Limeworks.
- Further work will be needed to see if lime of the required quality and consistency can be produced for the Malawi sugar industry, possibly from restarted production at the UML plant.
- Further work is needed on marketing lime, especially on its potential as an alternative to Portland cement.
- Investigation has started into gaining an entry onto markets in towns outside the two largest.
- A fuelwood plantation is to be established at Chenkumbi Hills.
- The possibility of re-establishing permanent lime production in the north, possibly at the Uliwa site, should be investigated.
- More encouragement should be given to new commercial producers of lime and lime-pozzolana cement in Malawi.

SECTION 3

Lime-pozzolana

INTRODUCTION

One potential application of lime which could extend its use as a building material is to combine it with a pozzolanic material. Pozzolanas are materials with an active form of silica (SiO_2) or alumina (Al_2O_3) which can react with lime to form additional cementitious components. Unlike lime, lime-pozzolana cement (LPC) hardens by chemical reaction with water. They also form a stronger cement than lime by itself and this cement sets less slowly. LPC has only a fraction of the strength of Portland cement so it is best suited for use with lightly loaded buildings and structures and those built with natural materials such as stone, earth and wood for which Portland cement can be considered too harsh and too strong.

Although not extensively used in East Africa, except perhaps in Ethiopia, research work of global significance has nevertheless been carried out on LPC in the region, chiefly at HABRI (Kenya), BRU (Tanzania) and the Geological Survey and Mines Department in Uganda, and papers are presented on the work carried out by each of these three institutions.

There is great potential in East Africa for increased utilization of LPC. There are extensive deposits of volcanic ash and rocks of moderate or high pozzolanic activity; in localized rice growing areas rice husk ash cement could be produced and in many cases local clays could be fired and then ground to a powder to give a reactive pozzolana. This gives an historical overview of developments in LPC production in East Africa as well as analysing the current potential for developing its application. It poses the question: why, despite increasing prices and growing shortages of other cementitious materials, is more use not made of LPC for building.

Tradition and history

LPC is a cement with ancient origins. Its first recorded use is in Greece several hundred years before the birth of Christ. Volcanic soil from the island of Santorin near to the Greek mainland was used as the pozzolana and the tradition of making and using LPC on the island still continues to the present day.

The Romans further developed LPC technology. Their main source of pozzolana was the red or purple volcanic tuff from the Bay of Naples region. The best material was considered to come from the area of Pozzoli or Pozzuoli, hence the pozzolana.

Despite the survival of some writing from the period, especially by Vitruvius, many aspects of the practice of producing and using LPC, developed by the Romans, were not documented and the knowledge was lost. Using LPC the Romans were able to build complex and daring structures which, even today, would be a challenge to engineers. Some of these structures and buildings, such as the aqueduct of Pont du Gard near Nîmes in France and the Pantheon in Rome, still survive more or less intact after some 2,000 years. LPC was also used by the Romans to build a harbour at Ostia demonstrating the durability of this material in harsh and aggressive environments.

Another material which the Romans used as a pozzolana was crushed underfired waste clay bricks, tiles and pottery. This type of pozzolana also has a long continuous history of use in India, where the technology developed independently and is known as surkhi. There are also traditions of using crushed fired clay as pozzolana in other countries, for example in Egypt and Indonesia.

Before the use of Ordinary Portland Cement (OPC) became widespread during the twentieth century, LPC was a commonly used cement in all types of construction. Hydraulic lime, a quite similar material to LPC (a type of lime which sets under water), was also used close to areas where the appropriate raw materials occurred. One famous example of a construction project for which only hydraulic lime and LPC cements were used was the Eddystone Lighthouse built in the 1750s off the south coast of England. The engineer for the project, John Smeaton, needed to have the confidence that the cement used would have specified setting, hardening, strength and durability properties. He conducted experiments with lime and pozzolanas from a number of sources before deciding on a mixture of Blue Lias hydraulic lime from the South of England and volcanic ash from Civita Vecchia in Italy.

In the past few decades considerable research has been done on properties and potential of using certain agricultural and industrial waste materials such as rice husks and pulverized fly ash (pfa). However, promotion activities have apparently achieved little success in gaining wider market acceptance of LPC at the expense of OPC.

Even in countries such as India, where LPC was used extensively until recently, OPC has become the most commonly used cement in many areas. Doubts also exist about whether production of LPC can be carried out as a sustainable commercial enterprise.

Comparative advantages and disadvantages between LPC and OPC

Advantages of LPC compared with OPC
Technical
- Good compatibility with traditional materials such as earth and soft stone, around 50% of the world's building stock.
- Much simpler production technologies.
- Most of the production plant can be made locally.
- Savings in energy consumption.
- Improved resistance to sewage and seawater.
- Lower heat of hydration hence offering more flexibility in mass concrete work.
- More attractive appearance.
- More tolerant to short-term exposure to moisture during storage and less likely to start to set quickly after mixing.
- less harsh and easier to spread and work in mortar and plaster applications.

Socio-economic
- Labour rather than capital intensive.
- Savings on foreign exchange.
- Lower investment costs
- Production responsive to local market needs.
- Plant maintenance can use skills which are locally available.
- Planning decisions at local rather than national level.
- Planning, installation and commissioning is less complex and time-consuming.
- Potential to exploit small-scale raw material reserves.
- Potential to use up some waste materials which are a problem to dispose of.
- Usually a lower-cost material, so more accessible for builders on low incomes.

Disadvantages of LPC compared with OPC
Technical
- Lower strength.
- Longer setting/hardening times.
- Behaviour can be less predictable.
- Often considered unsuitable for complex engineering structures.
- Generally unsuitable for concrete products, except blocks.
- Unsuitable for steel-reinforced concrete because of poor bond and lack of rust protection.
- Very difficult to exercise a high level of quality control in production.
- Higher cement dosages required to compensate for lower and less reliable strengths.
- Higher skill levels are generally required to use the material satisfactorily.

Socio-economic
- Lack of information and expertise in many areas.
- Lack of relevant standards and specifications in many areas.
- In some areas the traditional skills have been lost.
- The material needs to prove itself on the market, funders and customers could be very sceptical initially.
- Production at a profit might not be possible in some cases, in others profit margins would be low.
- Many potential customers perceive LPC to be inferior.
- The market for the product can be volatile rather than steady.

This introduction examines in more detail factors which have contributed to the decline of LPC and considers its future potential.

Technical aspects of LPC

LPC consists of a mixture of lime and a pozzolanic material. Research has shown that optimum mix proportions of pozzolana to lime are 2:1 or 3:1 by volume.

Pozzolanas are special forms of silicon or aluminium oxides, or a mixture of the two materials. Heating to between 500 and 800°C or vigorous grinding, or a combination of both, activates pozzolanic properties. Pozzolanas react chemically with lime and water to produce hydrated calcium and aluminium silicates. These materials are similar, but not identical, to those formed by the hydration of OPC.

Pozzolanas can also replace a proportion, normally less than 30%, of OPC in a blended Portland-pozzolana cement (PPC). In this case, lime liberated by the hydration reaction of OPC becomes available to react with the pozzolana. PPCs have similar hardening and strength characteristics to OPC and additionally confer some of the advantages of LPCs such as improved chemical resistance and lower heat of hydration. The use of PPCs is growing worldwide, especially in specialist construction applications. Their production is generally on an industrial scale and their quality control aspects are generally similar to those for OPC.

Compared with normal lime, LPC sets and hardens more rapidly, produces a stronger cement and, additionally, can set under water. Compared with OPC, however, LPC typically has only one fifth of the strength as well as longer setting and hardening times.

Table 38 compares specifications of OPC and LPC taken from Indian Standards. It should be noted that the strength values quoted are minimum values and it is possible to produce LPCs with strengths well above the minimum for LP40.

A number of materials exhibit pozzolanic qualities. These include pfa from coal burning power stations, slag waste from steelmaking, red mud waste from bauxite, a by-product of the aluminium industry, volcanic ash and some other types of volcanic rock, clay which has been heated to 500 to 800°C then finely ground, and ash from certain plant elements such as rice husks, rice straw and bagasse. Some of these materials, such as clay, require heating and grinding to make them pozzolanic, others, such as slag, only require grinding, and pfa and some volcanic ash can be used without further processing except, perhaps, sieving.

Some pozzolanas are more reactive than others and a number of chemical and physical tests to indicate pozzolanic behaviour have been developed. Regular testing of pozzolanas is important because their properties can be quite variable, even from the same source.

Table 38: Comparison of specifications of OPC and LPC

	OPC[1]	LPC[2]		
		LP40	LP20	LP7
Setting time (hrs)				
Min.	0.5	2	2	2
Max.	10	24	36	48
Compressive strength at 7 days (N/mm^2)	22	2	1	0.3

1. IS:269-1976 Ordinary and Low Heat Portland Cement (Third Revision)
2. IS:4098-1967 Lime-Pozzolana Mixture

The comparison on p. 109 gives some advantages and disadvantages of LPC as against OPC.

LPC in more recent times

Because LPC production has traditionally been an activity mainly associated with the informal sector, it has been impossible to obtain detailed statistics on the subject. However, there do seem to be general trends of gradual decline in the traditional production areas and lack of significant take up of the technology, with only a few exceptions, in areas where it has not traditionally been practised. In spite of this there is still continued interest in the subject from research organizations, non-government organizations, development agencies and even from potential commercial producers. More specific conclusions on the outlook for LPC can be arrived at by examining a few particular case studies.

Rice husk ash cement in India

India produces around 80 million tonnes of rice per year. Of this around 16 million tonnes would be husk which, if burned, would produce just over three million tonnes of ash. When mixed with lime potentially nearly five million tonnes of cement could be produced. This compares with annual OPC consumption of over 60 million tonnes.

In the 1970s and early 80s, India was experiencing severe shortages of OPC and a lot of effort was put into developing alternative cements. In the case of rice husk ash cement (rhac) there was close collaboration between research institutions, enterprise development and promotion organizations and entrepreneurs themselves. Probably several tens of producers set up, although exact figures are not known. The main use of the cement was in mortar or plaster work.

More recently there has been a considerable decline in rhac production, largely for two reasons – competition from new OPC production plants and increasing use of rice husks as fuel in domestic and industrial applications, so reducing its waste potential. Now there are probably only very few producers remaining.

In 1982 a rhac mortar typically cost 70% of an OPC mortar with similar properties. As the price differential between OPC and rhac has decreased in more recent times there has inevitably been a decrease in demand for rhac.

Lime-trass cement for concrete block production

Trass is a type of soft volcanic rock found in parts of West Java in Indonesia. Lime is also produced in the region. In this area there is still a flourishing industry producing concrete blocks from these two materials. These blocks are locally known as *batako*.

Production of *batako* is still being done using very simple techniques. The very soft rock can easily be dug out and crushed using manual methods. The trass is then dried and sieved to remove particles above 10mm across.

Slaked lime is then added either as powder or putty together with water, followed by a lengthy period of mixing with a shovel, or in a concrete mixer. Moulding of the blocks is done in a manually operated lever-type block press similar to the CINVA-Ram, or using a mechanized press.

Although recent information on production and use of *batako* is not available, it is likely that there has been little significant decline in its production and use. This would be for a number of reasons:

- production is simple and therefore costs are low
- the trass usually requires only sieving to remove larger particles although, if necessary, some grinding can also be carried out. The smaller trass particles react pozzolanically while the larger ones behave as aggregate.
- local knowledge of production and building with *batako* has developed over many years
- appropriate standards do exist
- lime production and trass deposits occur in close proximity

Although OPC is produced extensively in Indonesia it seems likely that *batako* production will still continue locally for the above reasons.

LPC in East Africa

East Africa is a region offering considerable potential for LPC production. Pozzolanas of volcanic origin would offer the greatest potential with Kenya, Uganda and Tanzania having extensive deposits of volcanic rock with some degree of pozzolanicity. A number of deposits of volcanic ash and tuff have been investigated for their pozzolanicity in all three countries and, in some cases, limestone deposits have also been located relatively close to pozzolana deposits.

Some 800,000 tonnes of rice paddy is produced annually in the countries of Kenya, Tanzania and Uganda, with Tanzania being by far the largest producer. If all the husks were used to produce ash for LPC then some 50,000 tonnes of LPC could be produced, but a realistic potential figure would be much lower. Although such a figure would not be greatly significant in terms of total consumption of cement in these countries, in a number of rice producing areas LPC based on rha could be significant locally especially for low cost building applications.

The only significant study on rhac cement in East Africa has been carried out by the Housing Research and Development Unit (HRDU), now the Housing and Building Research Institute (HABRI). This study identified a number of potential areas where Portland-pozzolana cement (PPC) or LPC based on rha would have significant local potential.

CAS can also report considerable interest in rhac production elsewhere in Africa, notably in Malawi and Nigeria, although it is not known if any production units have become operational.

It is probable that suitable soils or shales exist in Kenya, Tanzania or Uganda, for the production of burnt clay pozzolana, although little investigation on this has so far been carried out. A study in Burundi concluded that two types of kaolinitic clays gave active pozzolanas when heated to 750°C and that other types of clay produced somewhat less active pozzolanas.

Generally, in East Africa, indigenous production of LPC has not happened to any significant extent and projects by external agencies to introduce LPC technology have not met with much success. Two such projects are now cited.

Ruhengeri, Rwanda

Until the opening of a Portland cement (OPC) plant in south-west Rwanda in the mid-1980s, all OPC was imported. There was little significant production of lime or LPC within the country itself.

In the late 1970s, COOPIBO (Development co-operation of the International Building Companions), a Belgium-based NGO with extensive experience of working in Rwanda, formulated a proposal for setting up an LPC plant close to Ruhengeri. The main justification for the project was import substitution, with imported OPC levels reaching 20,000 tonnes and growing.

After a research and development phase, pilot trials and the signing of necessary agreements, production at PPCT (the Pozzolana, Lime and Peat Project) started in late 1983.

Lime was produced on site from nearby quarried limestone but the volcanic pozzolana was brought in from some 40km away along a tarmac road. Trials had shown that the most reactive pozzolana came from this source.

After firing in a continuously operated vertical shaft kiln the lime was mixed with pre-dried pozzolana before the mixture was pulverised in a vibrating ball mill for bagging and sale.

Technically, the project met most, but not all its objectives. The failings were that firing the lime kiln with local peat was not successful and planted eucalyptus had to be used instead, although peat could still be used for pozzolana drying. Also, the pozzolana proved to be insufficiently reactive with lime and a quantity of OPC was found to be necessary as accelerator. A final composition for the masonry cement was 66.2% pozzolana, 12.5% lime, 20% OPC and 1.3% minor constituents. Compressive strength tests indicated that the cement could reach a comparatively high figure of 10 N/mm^2 (MPa) after 28 days curing.

The plant reached close to full capacity in 1986 when production of LPC was six tonnes per day, the capacity being limited by the size of the ball mill.

As part of the project, tests were also done to determine optimum compositions of LPC based mixes for mortars and plasters and surveys on use of binders were carried out. It was found that LPC could be used satisfactorily in mortar and plaster work under building site conditions.

A study in 1987 indicated that even with the OPC plant starting in Rwanda, it would be viable to expand the PPCT plant to 6,000 tpa because the OPC plant would be unable to meet demand. However, the reality proved to be different. Sales declined severely in 1987 and production of LPC ceased in 1988.

The main factor in this decline was that the new OPC plant was able to make substantial price reductions. Relative to OPC the price of LPC masonry cement changed from 57% when the OPC plant first opened to 70% in 1988. Because a particular mix might require 50% more LPC than OPC, there was no cost saving to the buyer. Lack of familiarity with LPC and the fact that a higher level of workmanship is required to get satisfactory results also contributed to its decline. Another possible factor was lack of interest and support from influential bodies within Rwanda such as the Ministry of Industry and the Ministry of Works.

After the LPC plant closed down PPCT still continued to produce lime on a commercial basis for industrial, agricultural and building uses. However with the outbreak of civil war in Rwanda it is not known if the plant is still intact or how much repair would be required to restart production under more peaceful conditions.

It should be added that pozzolana similar in nature to that in Rwanda also occurs across the border in Uganda, and this has been investigated by the Department of Geological Survey and Mines in Uganda. The results of this study are described in a subsequent paper.

Arusha, Tanzania

A study in the early 1970s showed that in the Arusha region of Tanzania OPC was in short supply and lack of suitable raw materials meant that it could not be produced locally. However, it was estimated that locally produced LPC could meet 20 tonnes per day of the demand for cement in the region with production carried out in small-scale plants using relatively uncomplicated technologies.

In 1976, Intermediate Technology in collaboration with the Small Industries Development Organization (SIDO), a Tanzanian Government sponsored enterprise promotion organization, prepared proposals for a three year project to develop production and use of LPC in the Arusha region. Oxfam, a British-based charitable organization, agreed to contribute funding to the project.

The pozzolana used was volcanic ash from the Oldonyo Sambu area and the lime produced was slightly hydraulic. The pozzolanic activity of the ash was considered to be high, and higher than for the pozzolana in Rwanda, for example.

There were three components to the project:
(i) The building of the lime kiln and establishing LPC production.
(ii) Development of appropriate LPC based mortars, plasters and concrete blocks.
(iii) Construction of a store and demonstration building using blocks, mortars and plasters developed by the project.

At its peak the project was producing 20 tonnes of LPC per month and by the end of 1978 some 4000 LPC blocks had been produced. 6 to 10 persons were directly employed in LPC and block production.

The blocks were made with a CINVA-Ram lever type block press. The aggregate was a locally available pumice and so the blocks were of low density. Nevertheless their compressive strengths averaged 3 N/mm^2, more than adequate for most types of low rise buildings.

In 1980 the project was handed over to the Arusha Planning and Village Development Project (APVDP), a local counterpart organization. The Building Research Unit (BRU) in Tanzania also had some involvement in the project and used LPC blocks to build a clinic.

After 1980 production at the project site has stopped and to date has not resumed. This is despite interest in re-starting production shown by SIDO, the Tanzanian Army, BRU, the National Construction Council (NCC) in Tanzania, a local entrepreneur and Appropriate Technology International (ATI), an American-based development organization. In 1983 the United Nations Economic Commission for Africa (UNECA) proposed the establishment of a 'Sub-regional Development Demonstration Project for Lime Pozzolana Technology' with the Oldonyo Sambu project restarted and expanded to act as a demonstration cum training facility on LPC technology for the whole of Tanzania. However, no further progress has been made on these proposals.

Most of the production facilities at Oldonyo Sambu are still intact although the LPC block demonstration buildings are showing signs of neglect and decay.

As in the case of PPCT in Rwanda, the project was able to demonstrate technical soundness. However, as in Rwanda, there were other factors which eventually led to abandonment of the project and a complete lack of take up of the technology in the region, although in the case of Tanzania these factors were more varied and diverse. It has also to be added that in the case of Tanzania the project seemed to simply grind to a halt and no thorough evaluation was carried out at the end of it, so reasons why the project failed and lessons for future work were not well documented. However, the author considers the following factors of importance.

- There seemed to be a lack of planning, and the project just seemed to evolve according to short term needs and priorities rather than to a pre-defined plan. In particular little thought seems to have been given on how to disseminate the technology or how to involve the community with the project from the start.

- The project concentrated almost entirely on the technologies of LPC production and use. Economic, social and cultural concerns on LPC were not comprehensively addressed.

- The achieved production of the unit, at less than a tonne a day of LPC on average, was very low. Although a separate lime operation was also part of the project, the LPC operation by itself was probably too small to be commercially viable.

- There was a lack of infrastructure in the area so transport of raw materials and finished products presented a problem.

- There seems to have been little consultation carried out with the local counterpart organization who were charged with keeping the project in operation, on their management know-how, their interest in the project and on how they saw the project contributing to meeting their own objectives.

- Traditionally, house building has been carried out in the area by women, and as most of the local decision makers on the project were men they probably did not see continuation of the project as a priority.

- The project aimed to supply building materials to

a very localized area which was a very small market indeed. No attempt was made to supply the town of Arusha, for example, which is only some 32km away, or other nearby towns and villages.
- The Uganda-Tanzania war in the early 1980s disrupted production as well as affecting product demand.

As a postscript, interest was shown in the project by representatives from SIDA, a Swedish-based development organization, who in fact visited the project site in 1989. They were involved with a similar project in Ethiopia which included the building of 70 houses. They also carried out an evaluation of the Ethiopian project which apparently raised doubts about financial viability and the project was terminated.

Problems with development of LPC production and use

The above case studies show that projects on LPC are liable to fail unless a number of favourable circumstances apply, as in Indonesia. If the potential of LPC is to be realized worldwide it is important to consider possible constraints to its development in areas where conditions might not be so favourable.

Development organizations, especially those involved in building projects, might find promotion of LPC projects attractive for reasons such as utilization of local materials, local employment generation during the lifetime of the project, skills training for local people, and environmental benefits including waste utilization and reduced energy usage. However, the real test of long-term sustainability of LPC projects is their commercial viability which might be difficult to predict.

An entrepreneur intending to start up production of LPC would need to consider numerous factors all of which might mitigate against the commercial viability. These factors would include:

Technical:
- Not all pozzolanic materials are sufficiently active for use in cements and initial tests need to be done for pozzolanic properties.
- Pozzolanas are notorious for variation in properties from any particular source. This is particularly the case with volcanic ash.
- If heating or burning is required to activate pozzolanic properties, i.e. for rice husk ash and burned clay, this must be done under controlled conditions. The producer also needs to understand the processes involved.
- The mixing of the lime and pozzolana must be done very thoroughly.
- The pozzolana as well as the lime must be very finely powdered, preferably in a ball mill. It is rare, except perhaps in areas with a mining industry, that a workshop exists locally which could fabricate such a mill. An imported mill could cost $20,000, or more.
- With production using largely manual methods it is impossible to maintain the same quality control standards as in a more mechanised plant, implying the possibility of significant variation in the quality of the product.
- There may be a lack of standards, specifications or other documentation relevant to the materials of a particular area.
- The characteristics (eg. mix proportions, workability, rate of hardening etc.) of LPC based mixes are somewhat different from OPC based ones. Most builders would probably have trained only with OPC and so would need additional training in LPC applications.

Economic:
- Similar types of mixes will contain at least 1.5 more LPC than OPC due to the lower strength of LPC. The producer of LPC needs to be confident of selling the product at less than half the price of OPC to be confident of getting a foothold in the market.
- Production of less than a tonne of LPC per day would generate only very limited income to the producer. However, with higher levels of production it is important to ensure that the market is sufficiently large.
- A sceptical lending institution might not be convinced that LPC production would be viable.
- The viability of the project would be highly dependant on the supply and demand situation for cement in the area. In particular, if a new OPC plant starts up the LPC operation could collapse.
- In areas without an LPC tradition viability will only be achieved if a niche is found in the local market for the product, for example as an improved plastering material.

Cultural:
- If LPC has never been used in the area potential users will naturally be sceptical of the material. The producer must be prepared for a lengthy period of marketing and demonstration and, maybe, even training of builders before any sales are achieved.
- News of any bad experiences with LPC will quickly spread damaging its reputation. It is important to ensure that users of the material know where and how to use it and that any failures are thoroughly investigated.
- Because OPC might be considered more prestigious it might still be preferred despite its higher price.

Prospects for dissemination of LPC technology in East Africa

The greatest hurdle facing the potential producer of LPC is that a lot of time and effort needs to be spent on research and development, resourcing and marketing before production can be started and the product sold. Unless previous work has been done

by another organization, the potential producer would need to go through the following steps, not necessarily in this particular order:
- Identify raw materials, including the size and variability of any deposits.
- Prepare feasibility studies, including a market survey.
- Arrange the financing of the project.
- Arrange any necessary claims.
- Select production site and arrange for rent or purchase of land.
- Identify appropriate production technologies and equipment sources, and purchase the equipment.
- Set up the equipment and carry out production trials.
- Begin marketing of the product including, if necessary, preparation of instruction manuals, training of builders and putting up demonstration buildings.
- Begin production.
- Thoroughly investigate customers complaints, if any occur.
- Provide a back-up service for using the product on building sites.

It needs to be noted that only during the production phase will the producer be earning an income. During the previous phases the producer would need some other means of support.

If individual producers are expected to go through all these stages it is unlikely that sustainable production project will ever get off the ground.

However, if most of the non-productive phases could be taken out of the hands of the potential producer then the prospects for successful production should be improved. The non-productive phases could be taken over by one organization offering a back-up service to several producers and funded separately.

The back up organization could take on, much more effectively, activities such as geological investigation and materials testing, marketing, training of builders implementing changes to standards and building codes, and marketing at a targeted audience rather than at an individual level.

The back up organization could also critically analyse the supply and demand situation of binders in the region and identify particular niches in the binders market LPC could fill, if any. The social, developmental and environmental benefits which LPC can offer would additionally need to be considered. These benefits would include job creation potential, lower energy needs and less environmental damage than OPC production, and the potential for waste utilization as well as low investment costs. Most important of all organizations and individuals should only be encouraged to commit scarce development funds, own savings and other grants, loans or funds to LPC production if this has good potential for long-term sustainability.

References

1. Smith, R.G. Alternatives to OPC, Overseas Building Note OBN 198, Building Research Establishment, Watford, UK, March 1993.

2. Lea, F.M. *The Chemistry of Cement and Concrete*, 3rd Ed., Edward Arnold, London, 1970.

3. Smith, Ray. *Rice Husk Ash Cement: Progress in Development and Application*, IT Publications Ltd., London, 1984

4. Aksa, Z. 'Batako: A lime-based building material', UN Regional Housing Centre, Monograph no 24/2000/1983, Bandung, Indonesia.

5. Allen, W.J. 'Alternative Cement in Kenya and Tanzania'. A report for intermediate technology industrial services, internal report, March 1984.

6. Balu-Tabaaro, W. 'Research on Development of Alternative Cements Based on Lime Pozzolanas in Uganda for Use in Rural Housing', *Lime and Other Alternative Cements* (eds. Hill, N.; Holmes, S.; Mather, D.), pp. 105-118, IT Publications, London, 1992.

7. Tuts, R. 'Pre-feasibility Study on the Use of Rice Husk Ash as Cementitious Binder in Kenya', Housing Research and Development Unit, University of Nairobi, April 1990.

8. Olof Grane, M.P. Etude de la Fabrication de la Chaux, in UNIDO Project SM/BDI/73/011, Burundi, Final Report, UNIDO, Vienna, 1976.

9. Schilderman, T. 'The Use of Alternative Binders in Rwanda: A case study', in *Lime and Other Alternative Cements* (eds. Hill, N.; Holmes, S.; Mather, D.), pp. 218-228, IT Publications, London, 1992.

— Otto Ruskulis

HABRI's experience with pozzolanas in Kenya

B. Waswa

Among the main fields of research of the Housing and Building Research Institute (HABRI) of the University of Nairobi, is the development and dissemination of low cost appropriate building materials and technologies for housing in Kenya. After having concentrated on stabilized soil blocks for walling and fibre concrete roofing tiles for roofing, further research was started, focusing on binders. It should be noted that cement, which is the most widespread binder in Kenya, is sometimes unnecessarily used for applications where other binders could easily perform the necessary requirements at lower cost.

The following report deals with research carried out on rice husk ash (RHA) and naturally-occurring pozzolanas as replacements for cement. The research was carried out by HABRI in collaboration with the Department of Chemistry of the University of Nairobi as a HABRI/GTZ joint project.

Although RHA has been used in many parts of the world as a pozzolana (cement extender), its potential for such use in Kenya has remained largely untapped. There have been, however, recent attempts to investigate the economic and technical feasibility of utilizing RHA as a pozzolana in Kenya and the results so far show positive indications.

Rice husk ash

The potential application of a new binder largely depends on the availability, variety and relative costs of existing binders and the prevailing building practice.

Current estimates show that about 8,240 tonnes of rice husks are produced per year in Kenya. If anticipated expansion programmes are implemented, then by the turn of the century about 27,000 tonnes of rice husk will be accumulating as a waste annually. The husks have very low nutritional value for use as fodder and as they take very long to decay they are not suitable for composting manure. Since RHA has been found to be an excellent cement extender, it is important to convert the husk into ash for use in construction. Table 39 shows potential quantities of rice husk ash that will be available in Kenya.

By the turn of the century about 5,400 tonnes of RHA could be produced in Kenya. This represents about 1.3% of the national cement consumption. On a regional basis the figures are: Mwea can produce 3.2% of Central Province; Tana Delta 3.5% of Coast and Eastern Province, while Ahero can provide 1.2% of Nyanza and Western Provinces consumption of cement respectively. These are the areas in which rice is grown on a reasonable scale in the country. These figures are low, but if the product is competitive and cost-effective at a small production level, then it can be viable for a limited market. Map 6 shows the rice growing areas.

Table 39: Potential quantities of rice husk that will be available

Scheme	Actual formal	Actual informal	Proposed (formal)	Total
Mwea	1,200	200	874	2,194
Ahero	100	60	-	160
West-Kano	80	20	-	100
Bunyala	40	8	-	48
Tana Delta	-	-	2,912	2,912
Total	1,340	288	3,768	**5,396**

Equipment

The main equipment required for the production of the rice husk ash are a kiln and a ball mill. Also some laboratory equipment to measure the characteristic of the ground ash and the strength of the mortar made out of the ash is required. The technology can easily be handled by one skilled technician and two labourers.

The kiln at HABRI was built of burnt bricks and a galvanized iron sheet chimney. The platform on which the kiln is built is made of concrete. The kiln is essentially a modification of a kiln developed by the Pakistan Council for Scientific and Industrial Research. The height of the kiln including chimney is 3.32m, with a cross-sectional area of approximately 1.5m and an effective operating capacity of 1.8m^3. Figure 10 shows the kiln at HABRI.

The ball mill was fabricated by a Nairobi Metal Engineering firm and installed at the HABRI workshop. It consists of a cylindrical metal drum measuring 0.5m in diameter and 0.5m long. The grinding media steel balls are of various sizes ranging from 12 to 25 mm in diameter. The mill is driven by a motor of 4 kW and it revolves at 40 rpm. The capacity of the ball mill is about 27 kg of ash per batch. For commercial plant it has been calculated that a ball mill with a bigger cylinder (1.5 x 1.0m diameter) with a capacity of 325 kg of ash is preferred. Figure 11 shows the ball mill at HABRI.

Processing procedure

Six production procedures have been reported which can be used in producing RHA:

1. Controlled burning, grinding and addition of lime for use as lime-pozzolana binder.
2. Controlled burning of the husks, grinding the ash and mixing the ash with cement for use as Portland-pozzolana binder.
3. Heap burning, grinding the ash together with lime to produce lime-pozzolana binder.
4. Heap burning, grinding together with cement to produce Portland-pozzolana binder.
5. Burn husk and lime sludge balls together, grinding the resulting ash to produce lime-pozzolana binder.
6. Burn husk and clay balls together, grinding the resulting ash to give Portland-pozzolana binder.

Table 40: Blaine test on the effect of grinding time (hrs) on the fineness (cm^2/g) of RHA

Batch No.	1	2	3	4	5
1	1898	2001	2982	3867	4191
2	1768	2148	3193	3846	4185
3	1994	2064	3210	3699	4096
4	1768	2179	3211	3519	4004
Average	1857	2098	3149	3733	4119

Out of these processes, Option 2 was chosen and used at HABRI for the following reasons:

- Controlled burning increases the reactivity of the ash and does not require a major capital investment.
- In order to produce lime from lime sludge and reactive clay from clay balls higher temperatures are required than the optimum temperatures for rice husk ash, therefore lime sludge and clay ash (Options 5 and 6) were excluded.
- Lime for building is far less popular than cement in most parts of Kenya because of its high cost at the time and relatively low quality for building applications; therefore lime applications (Option 1) were given lower preference than use as a cement extender.

The procedure chosen therefore was that of burning the husks in a controlled manner (temperature less than 700°C) in a special incinerator. The control is achieved by variable air inlets.

Rice husks were collected from Mwea and Kisumu and transported to Nairobi in a truck. The Rice mills did not charge for the husks. However, in the subsequent cost calculations, it is assumed that the husks have a nominal value of KSh 20 per tonne.

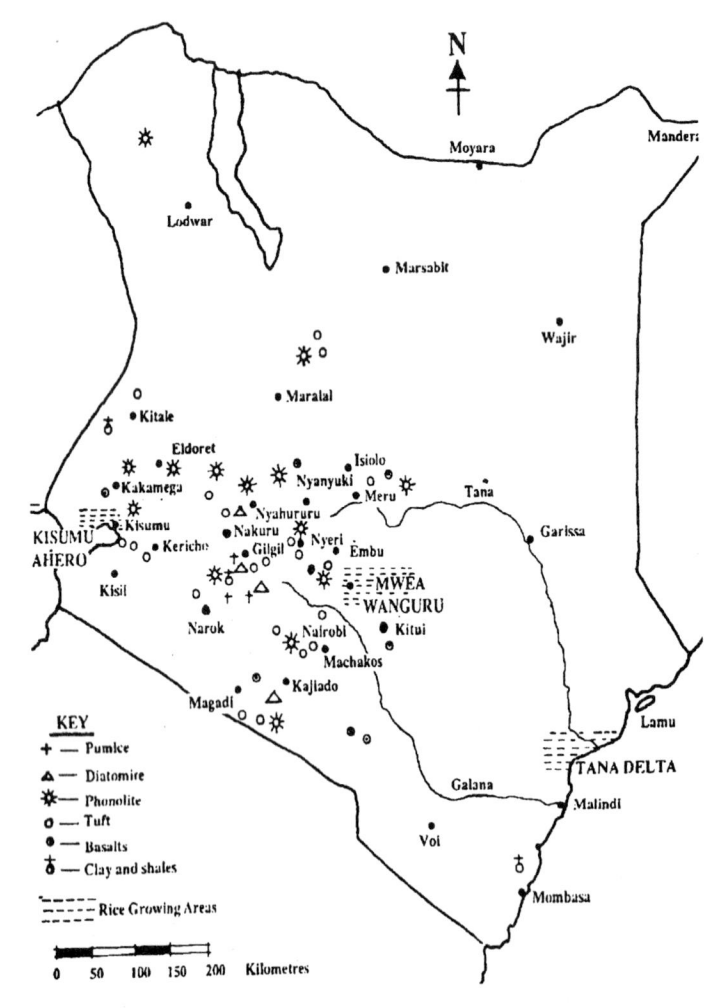

Map 6: Locations of rice-growing areas and natural pozzolana deposits in Kenya

Calcination

About 8 bags of husks each weighing 22-25kg were needed to fill the kiln with a light compacting effort. A fire was then started at the bottom of the kiln with small pieces of wood and paper. After this no other fuel was needed for the combustion of the husks. Temperatures were recorded with a thermocouple with a digital reading device.

Table 41: Chemical composition of RHA from Mwea and Ahero/Kisumu

Constituent	% composition Mwea	Ahero/Kisumu
SiO_2	85.00	89.44
Fe_2O_2	0.23	0.41
Al_2O_3	0.32	0.46
CaO	0.67	0.58
MgO	0.43	0.42
Na_2O	0.07	0.47
Mn_2O_3	0.12	0.14
K_2O	1.24	1.35
TiO_2	-	-
P_2O_5	0.71	1.55
Loss on ignition	6.93	3.66

The average temperatures over the peak period (between 20 and 30 hours of burning) were around 540°C. The highest temperatures for the different reading points varied between 580°C and 660°C. The husk ash was removed after 45 hours when the burning was completed and then allowed to cool down to ambient temperature. It should be noted that if the ash were to cool down in the kiln, the burning cycle would take longer. Ash with varying carbon content (and therefore different colours from grey to dark grey) was obtained, depending mainly on the ventilation available and on the time of removing the ash.

Grinding

After burning, a measured volume of the ash was placed into the ball mill and made to grind for six hours. Samples were taken after every hour to monitor the increasing fineness of the ground ash. The milled ash was then packed into nylon gummy bags kept in a dry place and transported to the project sites in Makueni and Maseno. The effect of varying RHA grinding time on the fineness (cm^2/g) of RHA is shown in Table 40.

Chemical analysis of the ash is shown in Table 41. X-ray diffraction analysis showed that the silica in the ash is substantially amorphous.

Figure 10: An experimental kiln at HABRI

Application tests

To test the response of RHA-cement to environmental conditions different formulations, based on laboratory tests and a literature survey, were used as mortars at HABRI workshops and different buildings in two districts: Makueni (Machakos) and Maseno (Kisumu).

Table 42 shows the preliminary tests that were carried out using various mixing proportions of OPC and RHA.

At the HABRI workshop the mortar mixes were applied as plasters on external walls at various points. The recipient walls are made of stabilized soil blocks. The ratios used were 50:50, 70:30 of OPC:RHA.

At Maseno Youth Polytechnic, mortar mix ratios 20:80, 30:70, 40:60, & 50:50 (RHA:OPC) were applied at different portions on external walls. The

Table 42: Some technical characteristics of the RHA mixes

% OPC	% RHA	Compressive strength (mn/m) 3 days	7 days	28 days	Setting time (min) Initial	Final	Water/cement ratio	Fineness (cm^2/g)
100	0	29	34.4	41.7	180	267	0.40	380
95	5	41	51	57	115	242	0.405	495
90	10	41	52	55.5	109	239	0.422	390
80	20	42	48.5	53.7	109	185	0.454	463.7
70	30	31.3	39.7	46.7	94	154	0.459	430.5
60	40	29	36.7	42.3	100	194	0.605	466.2
50	50	23.7	25	30	26	84	0.562	423.8

Table 43: Strength of RHA plasters (28 days) used in Makueni

Plaster type:	1	2	3	4	5	6	7	8	9
Cement	2	1	1	1	1.5	1.5	1.5	0.5	0.5
Lime	0	0	1	1	0	0	1	0.5	1
Ash	0	1	1	0	1	0.5	1	1	2
Sand	6	6	6	6	6	6	6	6	6
Total volume	8	8	9	8	8.5	8	9.5	8	9.5
% Cost	100	67	91	93	83	84	105	64	105
Rupture str. kn/mm	4.6	4.2	1.4	1.1	4.6	4.2	7.2	3.5	4.9
Compr. str. n/mm^2	2.7	2.7	1.9	1.8	2.8	2.4	2.2	1.5	1.8
Relative rupture strength	100	92	30	23	100	92	153	76	107
Relative compr. strength	100	102	70	67	103	91	82	57	60

40:60 mix appears dark in colour due to the colour of the ash. In one of the classrooms the floor finish at the door entrance was made using the mixing ratio cement:RHA: sand, of 1:2:6 respectively.

More detailed application tests were done in Makueni, Machakos where a homestead with various buildings was used as the project site. The materials used were cement from Athi River, lime from Mombasa, RHA produced at HABRI with husks from Mwea and sand from Machakos.

The conventional 1:6 mix (cement:sand) for plaster was gradually modified by reducing the amount of cement and adding various amounts (measured by volumes) of RHA and lime. Attention was paid to the use of simple volumetric ratio between the different components, which are easy to measure in ordinary building site situations.

Mortars containing RHA were applied on various building components related to wall and floor finishes, both inside and outside. In some cases, complete walls were plastered with different mixes. In other instances, different mixes were used, so as to compare the performance of the different plasters in similar conditions and with the same exposure to climatic conditions.

Table 43 shows the mortar compositions that were used in Makueni. The recipient walls in this case are made of burnt bricks.

Fineness of the ash

The fineness of the ash is an important characteristic in that it allows comparisons with the fineness of the cement. It is known that increases in fineness of a binder increase its reactivity resulting in increased strength of mortar. The fineness is measured with the balance apparatus for air permeability, which estimates the surface area of a sample binder. The common standard of fineness of KS cement is 3550cm/g or 355m/kg. The effect of varying RHA grinding time on the fineness of RHA is shown in Table 43. From Table 43, it can be noted that the RHA achieves the fineness of cement after a grinding period of between 3 to 4 hours. For the applications, the RHA was ground for 5 hours which gave 16% finer materials than OPC.

Since there is a direct relationship between fineness and strength, while the electricity requirements of grinding RHA considerably contribute to the overall cost of the product, it is essential to optimize the grinding period of the ash. The results in the table show that after four hours of grinding a higher fineness than OPC is achieved. The results appear to urge for proper balancing of fineness, grinding time and the related electricity consumption during grinding.

Appearance

The appearance of the RHA is determined largely by the amount of carbon in it. Very light grey ash was obtained when RHA was allowed to cool to ambient temperature in the kiln. This however required a much longer time than when the ash was removed while still hot and allowed to cool in the open. Dark grey ash was obtained when the ash was allowed to cool in the open.

Workability

Figure 11: A ball mill at HABRI

The workability of the mortar affects the adhesion to the recipient wall and the time spent on plastering. It was noted that slightly more water was needed for the mixtures with large amounts of RHA to achieve adequate workability.

Drying Pattern

The drying pattern and drying speed of some of the plasters could give information on the need for consistent watering during the first days after plastering. Table 44 shows the speed of drying of some selected plasters. It can be seen that the plaster containing RHA and cement only (2) had a faster drying pattern than the conventional plaster (1) RHA cement mortar therefore appeared to lose moisture faster than the other three mixes. This observation, which is in line with known properties of pozzolanic binders, emphasizes the need for regular, frequent watering of RHA cement applications.

Durability

The durability of the plasters can only be really measured over longer periods. Important aspects include resistance to impact, brushing and weather. The plasters applied to the exterior were exposed to rain. However, during the inspection visit in June 1994 after 2 years, there were no evident indications of wearing of the plasters containing RHA due to weather action.

Table 44: Drying pattern of selected mortars

Time (mins)	OPC	OPC/RHA	OPC/lime/RHA	OPC/lime
0	100	100	100	100
20	95	80	100	100
40	80	65	80	75
60	50	40	55	60
100	35	30	40	50
120	15	5	15	25
140	5	0	10	15
160	0		0	5
180				0

Rupture and compressive strength of prisms

The mixes which were used on site were also tested for their strength characteristics in the laboratory. In these tests the amount of sand was systematically reduced to half (from 6 volumes to 3 volumes) in order to obtain significant sensitive readings from the crushing machine. The dimensions of the prisms were 40 x 40 x 160 mm. First each prism was subjected to the rupture test after 28 days, after which the two resulting small prisms were compressed.

Referring to Table 43, plaster 2, where 50% of OPC was replaced by RHA (by volume), has rupture and compressive strength characteristics almost similar to conventional OPC mortar and a relative cost saving of 33%. The mortars used in plaster 5 and 6 (with 40 and 25% replacement of OPC by RHA respectively) and with relative cost savings of 17 and 16% were of similar rupture and compressive strength compared to the conventional OPC mortar. The other mortars (all of them containing lime) showed significant lowering of either compressive or rupture strength and minimal cost savings relative to OPC mortar.

Table 45: Description and reserves (million tonnes) **of selected natural pozzolanas**

Name	Location	Description	Colour	Reserves
Pumice	Longonot	Pebbles	Grey	62.4
Diatomite	Karianduzi	Soft rock	White	2
Phonolite	Embakasi	Dust	Grey	5.1
Yellow tuff	Njiri	Soft stone	Yellow	2.5
Athi tuff	Athi River	Soft stone	Pink/grey	23

Source: Tuts, R (1991)

From the viewpoint of the ease of mixing and strength characteristics of the mortar, the preferred mixtures in order of desirability are 2, 5, 6 and 8.

Natural pozzolanas

Kenya has abundant reserves of natural pozzolanic materials in various sites all over the country. The main types of pozzolanic materials found in Kenya are pumices, diatomites, phonolites and tuffs. They are largely lying unexploited. HABRI conducted limited tests on samples of pumice from Longonot, phonolite from Embakasi (Nairobi), diatomites from Karianduzi and yellow tuff from Athi River. These materials were found suitable as pozzolanas to varying degrees (50%-15%) of cement replacement without adversely affecting important concrete properties such as strength, workability and setting time. Table 45 gives some characteristics of some of the potential natural pozzolanas.

The combined values of $SiO_2 + AL_2O_3 + Fe_2O_3$ for these pozzolanas vary between 53% and 84% (Table 46). The concrete cube's strength, as compared to OPC only as binder, shows that 3 of the 5 pozzolanas have systematically higher strengths than OPC when mixed with a 15:85 ratio. For the 30:70 ratio, still one sample gets higher values than OPC. The diatomite sample could only be tested up to 25% because with richer diatomite mixes, the water requirement becomes very high. No tests have yet been carried out on cements of natural pozzolanas mixed with lime.

The use of natural pozzolanas can be pursued on a large-scale because of their abundance in nature but one should expect more problems in terms of quality of the final product. This is because of the

Table 46: Characteristics of selected natural pozzolanas

Name	Chemical analysis (ground %)[1]	Density (OPC only) kg/m	Concrete Cube 15% pozzolanas 28 days	60 days	Compressive strength (%) 30% pozzolanas 28 days	60 days
Pumice	32	2475	92	94	79	83
Diatomite	84	2020	109	114	-	-
Phonolite	57	2705	88	93	77	82
Yellow tuff	53	2240	120	124	102	106
Athi tuff	76	2430	100	100	82	89

1. ($SiO_2+Al_2O_3+Fe_2O_3$) *Source:* Tuts, R (1991)

enormous variations in reaction of the natural pozzolana available in Kenya.

Proposed follow-up

With these successful initial investigations HABRI intends to carry out an expanded research programme on RHA cement. The research aims to:

- Set up a regular production unit for RHA at HABRI, University of Nairobi.
- Extend the production unit to include burnt clay pozzolana.
- Establish the quality of conventional binder that can be replaced by the RHA or burnt clay.
- Explore markets for the products.
- Introduce small-scale investors to the technology.
- Assist in selecting of pioneer production sites depending on raw materials availability
- Recommend standards for production and use of RHA and burnt clay pozzolanas to the relevant standardization authority.

References:

1. Tuts, R., 'Potentialities and Constraints for using pozzolanas as alternative binders in Kenya'. Paper presented at the 1st International symposium on Lime and other alternative cements. Stoneleigh, ITDG, 1991

2. Tuts, R., 'Prefeasibility study on the use of rice husk ash as a cementations binder in Kenya'. HRDU/HABRI 1990

3. Hammond, A.A., "Research on rice husk ash binders in low-cost housing technology," HRDU/HABRI 1991

4. Mbindyo, J.K.N, Kamau G.N and Tuts, R, 'Recycling of rice husk wastes for use as cement replacement material in Kenya'. Paper presented at World Conference on Philosophy. Environment and Development session Nairobi 1991

5. Mbindyo J.K.N., 'Chemical characteristics of rice husk ash and its applications' MSc. thesis, University of Nairobi, 1992

6. Tuts, R and Mbindyo J.K.N., 'Application tests on Rice husk ash binders'. HRDU/HABRI 1992

Pozzolanic cements in Uganda

W. Balu-Tabaaro

Portland cement has a clearly defined role as binder in fulfilling requirements for high strength application. In Uganda as in many other third world countries, Portland cement is unnecessarily being used in low strength applications: foundation concrete, plasters, mortars and soil stabilization.

This wrong application of Portland cement is not only unnecessarily costly but technically defective. Portland cement mortars are at times too harsh and lack plasticity, and do not spread easily resulting in partially filled joints. This leads to lack of intimate contact between the mortar and masonry units.

The degree to which Portland cement is wrongly used has reached alarming proportions and it is estimated that only 20% of the worldwide use of cement requires the strengths of Portland cement. Building technologists have found that a 0.55 MPa mortar can support a four storey building. Lack of alternatives to Portland cement has created this untenable state of affairs.

Because of the need to provide cheaper cement, the Department of Geological Survey and Mines approached the International Development Research Centre of Canada, IDRC, who agreed to fund a research programme on alternative binders based on pozzolans. This research programme, which took 4 years, used volcanic ash from Kisoro and Bunyaruguru, and reject clay bricks and various lime products.

Pozzolana cements

Under the Pozzolan Housing Project the main objective was to come up with acceptable and affordable cements for low cost housing construction. Several pozzolan cements based on volcanic ash, tuffs, burnt clay rejects, lime and ordinary Portland cement have been produced, tested in the laboratory and also used in the construction of a model house. Mortar strengths of 0.48-2.70 MPa were realized for lime-pozzolan cements, while 28 days strengths ranging from 13.20-29.0 have been achieved for Portland-pozzolan and Portland-lime-pozzolan cements.

The power consumption for production of pozzolan cement was estimated at 125 kWh/tonne using a 3" x 3" ball mill in the laboratory. It is anticipated that a continuous small plant would have lowered these values.

Raw materials base

Alternative affordable and appropriate cements can be manufactured through use of pozzolanic materials blended with lime, Portland cement or a combination of the three. Most pozzolans do exist in the country as idle resources or even, as waste materials, presenting a disposal problem.

Pozzolanas

The types available in Uganda are outlined below.
Natural pozzolanas: This grouping includes volcanic ash, scoria, pumice and tuffs occurring extensively in Kisoro-Kabale areas and to a lesser extent in Bunyaruguru and Fort Portal areas. Millions of tonnes of such materials are known to exist.

Shales and diatomaceous earth occur extensively in various areas:

- Toro-Ankole area as light grey sediments, mainly ashy shales and mudstones. The reserves have not been estimated.
- Mt. Elgon area, they exist as sandstones and fine grained black/dark grey ashy shales and mudstones.
- In Pakwach area the reserves of diatomaceous earths are estimated at 100,000 tonnes.

Artificial pozzolanas: These comprise of pozzolans of agro-waste ash and industrial wastes. The former include coffee husk ash, ground nuts shell ash, bagasse ash and others. It is estimated that the amounts of the agricultural waste are as follows:
Coffee husk — 650,000 tonnes
Bagasse husk — 850,000 tonnes (UNIDO estimates).

Soft fired clays, clay reject products and fly ash are also pozzolanic.
Chemical analysis of pozzolanas: Chemical analysis was done for Kisoro volcanic ash, Bunyaruguru tuffs and Kajjansi clay reject bricks. A pozzolan to be used in cement production should have a combined percentage weight of silica and alumina greater or equal to sixty. Samples meeting the above criterion yielded the best results.

Lime

Lime is produced at Hima and Tororo near the cement factories using the same limestone resources. The reserves were estimated at 23 million and 6 million tonnes of limestone for Hima and Tororo respectively (1969). Other important locations are Dura and Muhokya in Kasese District where a number of small-scale works are reported. The Dura deposits are secondary limestone derived from lime leached from calcareous volcanic tuffs and carbonate springs. The reserves are estimated at 1.2-1.5 million tonnes. The reserves at Muhokya have never been adequately assessed but a deposit of at least ¾-million tonnes has been reported.

Another limestone deposit occurs on Kaku river, 16 km from Kisoro town, there the reserves are estimated at 1.75 million tonnes. There is hence sufficient limestone to be exploited for the production of pozzolan cement.

Available lime index: Most Ugandan small-scale lime burners produce low-grade lime due to poor kiln designs. A rapid sucrose test was done on hydrate and lime samples. The results are summarized below.

Sample No.	Origin	Available Ca(OH)$_2$ (%)
1	Kenya Lime	92.0
2	Equator Lime Kasese	81.2
3	Equator Lime Kasese	76.6
4	NEC Lime	64.7
5	Kisoro District Adm. Lime Works	38.3
6	Muhokya Lime	38.0

Limes No. 5 and No. 6 contained a lot of unburnt limestone hence have very low amounts of free lime available for pozzolanic cement reaction.

Application of pozzolana cements

The various cements produced were tested in the laboratory as well as in the field and observed for periods ranging from 6 months to 2 years. The ensuing discussion is based on the laboratory and field observations.

Table 47: Lime-pozzolan-Portland mortar strength (compressive strength MPa – 28 days)

Pozzolan type	Lime-pozzolan cement	Lime-pozzolan + 10% OPC	Lime-pozzolan + 30% OPC
Kisoro volcanic ash	0.93	7.03	15.35
Kajansi burnt clay	0.48	7.78	15.57
Bunyaruguru volcanic tuffs	2.70	-	22.23

Lime-pozzolana cements: These cements were made with a composition of 25% lime and 75% ground pozzolans. They exhibited low mortar compressive strengths ranging from 0.48-2.70 MPa. Lime-pozzolan cements showed excellent workability when used to bond masonry units, this applied to both burnt clay bricks and stabilized soil blocks. Slightly longer setting times did not adversely affect the progress of construction work, as observed from model house construction.

Lime-pozzolan-Portland cements: Addition of Portland cement to lime-pozzolan cements in the range of 10-30% enhanced the strength of these cements. This effect is illustrated in Table 47.

The lime-pozzolan cements with 30% OPC have been used in slabs to simulate foundation concretes. Strengths of 5.9-11.0 MPa have been registered in 1:3:4 concrete mixes (cement:sand:aggregate) and water cement ratios (W/C) of 0.65. Plasters done using this mix provided excellent workability and nice finishes in the field tests.

Portland-pozzolana cements

Portland cement replacement by a pozzolan were done up to 50%. The mortar strengths for 28 days varied from 38.4 for Portland cement to 13.5 and 20.94 for Kisoro and Bunyaruguru tuffs at 50% Portland replacement by pozzolan respectively.

The strengths attained for blended ordinary Portland cement (OPC) to pozzolan (OPC/POZZ) 80/20 and 70/30 were found to be acceptable for general concrete works. (Designation ASTM C: 595-86). The blend OPC/POZZ 60/40 can be classified for general concrete construction where high strength at early stages are not required. Blend OPC/POZZ. 50/50 can be classified as a type of masonry cement for use in binding and rendering works and other simple construction chores involving low load bearing members.

Estimation of pozzolana-cement production costs

The cost of production of lime-pozzolan and blended cements were estimated assuming a 3 x 3 ft, 20 hp Denver ball mill installed at the production site. Kisoro being the furthest source of raw material was selected to serve as an indicator of the cost of production, with the production site assumed as Entebbe.

Operating parameters:

- 300 days working year was assumed.
- Plant capacity 2.25 tonnes of cement per 8 hour working shift;
- Daily volcanic ash requirements are 1.7 tonnes, hence annual total would be 510 tonnes;
- Annual lime requirements 168 tonnes;
- Cost of volcanic ash Sh 4,000/tonne (US$4) from Kisoro
- Cost of lime Sh 160,000/tonne ($160/tonne)
- Transportation of volcanic ash by 10 trucks hired at Sh 350,000 per trip ($350); 51 trips are required (Kisoro-Entebbe).

If the production was based at Kisoro the cost will definitely be much lower since transport costs would be almost eliminated. This would make the cement available at cheaper cost for housing constructions.

Conclusion

The following conclusions can be drawn from the field and laboratory experimental work performed on Ugandan pozzolans.

The volcanic ash and tuffs sampled in Kisoro, Kabale and Bunyaruguru proved to be reactive when combined with high quality lime. Similar results were exhibited by a reject burnt clay products from Uganda Clays at Kajansi. Thus these pozzolans can be used in the production of lime-pozzolan cement or as admixtures to ordinary Portland cements at cheaper cost.

Lime-pozzolan cements

Lime-Pozzolan cement produced using Kisoro volcanic ash, Bunyaruguru tuffs and reject burnt clay products provided excellent workability as binders for masonry units. Mortar compressive strengths 0.48-2.70 Mpa were achieved by these cements. Addition of 10-30% Portland cement to lime-pozzolan enhanced the strength development.

Portland-pozzolan cements

Blended cements produced through utilization of the local pozzolans with Portland cement achieved sufficient strengths to be utilized in general concrete construction in the low and middle cost housing schemes.

The power consumption for production of pozzolan cement was estimated at 12.5 kWh/tonne of cement using a laboratory mill.

Savings of 30-50% in cost of cement are achieved when blended cements and lime-pozzolan cements are produced for low-cost housing compared to Portland cement.

The development of pozzolanic cement production at Kisoro and Muhokya will enhance employment opportunities and increase incomes in these rural areas and help to improve the housing stock.

Recommendations

It is recommended that production of quality lime be given serious attention. Most kilns used and burning techniques are poor. These result in incomplete burning of limestone rendering the products inferior in grade.

Pilot plant production needs to be set up in Kisoro and Bunyaruguru to continue the assessment of lime-pozzolan cement and related products.

Lime-pozzolan cement can be used as binder for the masonry units. In case of foundation and plastering 30% Portland cement should be added to enhance early strength development.

Research of the Building Research Unit (BRU), Tanzania

Livin Henry Mosha

The views expressed in the paper are those of the author and do not necessarily reflect those of the BRU.

The Government of Tanzania in 1970 decided to establish a National Housing and Building Research Unit under the Housing Division, Ministry of Lands, Housing and Urban Development. BRU started its obligations in 1971. BRU has a mandate to carry out research and development, and information dissemination on building and housing technologies and issues relevant to Tanzania. Among the areas on which BRU has been working have been appropriate building materials, housing and shelter requirements and designs, building economics and management, and the regulatory framework for appropriate building.

BRU activities

Most of the activities BRU has undertaken can be grouped under four distinct headings:

(i) *Literature publication*: to date, more than 120 publications on various aspects of building construction and building materials have been produced.

(ii) *Seminars:* BRU have so far conducted a limited number of seminars to both urban and rural dwellers with the focus on innovative building materials and technologies.

(iii) *Media:* radio programs and articles in newspapers on outcomes of BRU projects.

(iv) *Demonstration houses:* Far and above all the above three strategies, constructing demonstration houses was considered to be the most appropriate for beneficiaries to see, learn and adopt innovations. About 14 demonstration houses were constructed by BRU in various locations incorporating many original and innovative designs and features in foundation, wall and roof construction.

This strategy has raised particular concerns, especially on the appropriateness and relevance of the research and dissemination work to the real housing needs of communities. In particular the dissemination efforts have been considered to reach only a few people as lack of funding has prevented the preparation of targeted publications and the construction of more demonstration buildings designed specifically for local conditions. Tanzania is a country with a diverse range of environments and one particular design would not always be appropriate in all areas. Additionally the introduction of appropriate materials and designs is hampered by inflexible and irrelevant building codes and regulations and a professional bias against materials such as earth and lime.

BRU work on binders

BRU has produced a number of publications on binders on subjects as diverse as pozzolanic cements, plasters and plastering techniques, location of raw materials for alternative cements production, and the design of simple lime kilns.

In addition BRU was involved with a lime-pozzolana cement production project at Oldonyo-Sambu near Arusha and built a demonstration clinic there. This project is described below.

Use of lime and pozzolime in Oldonyo Sambu, Arusha

ITDG has been involved with SIDO in a lime-pozzolana Project at Oldonyo Sambu in Arumeru district, which although producing a promising material, failed in the long run. BRU has also been involved in promotion of use of pozzolana and lime (pozzolime) in Oldonyo Sambu. BRU started a demonstration building early 1978 in collaboration with SIDO-Arusha. The objectives of the project were:

- to raise villagers awareness on how to use the local building materials available.
- to find out total building costs of using pozzolime.

The pozzolana was volcanic ash from the Mt. Meru volcanic crater. Limestone for lime production was bought in from deposits nearby and lime was produced on site. The lime pozzolana cement was used to produce concrete blocks and for plastering mixes, and both these materials were used to build the clinic annexe referred to below.

Oldonyo Sambu demonstration house: This is an annex to a dispensary, with the total floor area of $54m^2$. The project cost TSh 25,313 ($3038). This works out at TSh 468 ($56.27) per m^2. The 1977 Exchange rate was $1 to TSh 8.337 (Source: Bank of Tanzania, Economic and Operations Report, 1977).

Through interviews made in August 1994, it was noted that the impact of this project to villagers is negligible. Probably factors such as a reluctance to use a new and unfamiliar material for construction, inadequate dissemination activities aimed at builders in the area and, especially, the relatively high cost of construction jeopardised any attempts to use lime-pozzolana cement for building elsewhere in the area.

However, there were a few elements of positive impact from the project, such as:
- In 1984, Arumeru district council requested BRU to conduct further research on local building materials in the district.
- In 1988 Zanzibar Officials visited the project so that they could see and learn how pozzolime was used.
- Currently, there are two groups of artisans namely Umoja wa Mafundi Oldonyo Sambu (UMO) and Ufundi Lemong'o Society (ULS) which are very eager to learn about low cost building materials and associated skills. They are so interested to promote pozzolime but their major problem is CAPITAL.

References:

Bank of Tanzania, "Economic and Operations Report", Dar es Salaam, 1977.

BRU working file, "Demonstration House on Pozzolana at Oldonyo Sambu", Dar es Salaam (1978-1994).

Edvarsen, K and Hegdal, B. "Rural Housing in Tanzania: Report on a pre-study", Dar es Salaam, 1972.

Lyamuya, P. "The Rural Built Environment in Tanzania: A Study of rural settlements and housing conditions with critical review of past policies and programmes, and a proposal for an alternative approach based on case studies in Uchagga, Leuven, 1990".

SECTION 4

The use of binders in low-income housing

Theo Schilderman

When comparing dwellings where binders have been used with those that have not, the importance of binders in improving the living conditions of the inhabitants of East Africa only becomes too clear. Not so long ago, the average lifespan of a house in Tanzania stood at 7 years, which is not astonishing considering that traditional housing is usually built with earth and organic materials such as poles and thatch, which do not last long when unprotected from the climate and insects. A proper application of binders can provide protection against humidity, erosion and cracking and make it more difficult for insects to harm a house, and so increase its durability considerably.

This chapter is about what motivates or constrains low-income people to use binders in housing. Important factors in this context are affordability, availability, cultural acceptability and awareness.

Three in-depth surveys were done to create a picture of the use of binders in low-income housing in Kenya, Uganda and Tanzania. At the end, Zanzibar is briefly treated as a special case with a long tradition of using one specific binder, lime, which is representative for some still surviving building technologies in areas along the coast and on the islands.

The surveys were meant to get some insight into the use of binders in housing, not to present a representative sample. In each country, a number (5-7) of settlements were selected, not at random but following established criteria; the idea was to get a fair mix of locations that were either rural or urban, close to or away from commercial binder production, and with easy or difficult access. In each location, a sufficient number of houses were surveyed where binders had been used, and these were compared with a sample of houses where no binders had been used. From the figures in this chapter, it may appear as if binder use is predominant, but in reality it is not. There are far more houses where no binders have been used, although the percentage of binder use is slowly increasing, as can be noticed from subsequent censuses in Tanzania, for instance.

In their use of binders in housing, the East African countries have quite a lot in common, though differences can also be noticed, both between and within countries. An example of the latter is the much wider use of lime on the islands and in some coastal locations, mentioned above.

The choice of binders in Kenya is influenced by both cost and availability of the various alternatives, as well as the additional inputs they require both in terms of materials such as aggregates and skills; cultural acceptability is another important factor. Cement is expensive, but relatively easily available; it is mainly used in formal construction. Lime is not a lot cheaper, and its use is limited by a lack of confidence of its users (even the lime plant at Koru mixes it with cement for its staff housing) as well as ignorance of its use on the part of the builders. But even the use of cement, though much better known, could be improved by better mixing, curing etc. to reduce waste and costs. It is estimated that cement and lime together are used in less than 10% of informal housing. Indigenous binders such as cowdung and ash as well as special earths are still often used, but have a lower status; they are often not very durable. Their use is coming under threat of lesser availability of the materials and a gradual loss of the traditional skills involved in some areas. Blended binders, often using pozzolanas, are slowly entering the Kenyan market, but still relatively unknown; although cheaper than cement and lime, they may still not be cheap enough for the lower-income groups. Their hopes may lie in improvements of the indigenous binders, e.g. by stabilization with lime.

In Uganda, the picture is not radically different. Cement is expensive and sometimes hard to get. Lime is not considered to be an adequate binder on its own, and mixed with cement. The preference for cement is enhanced by the training artisans receive in technical schools, which focuses entirely on cement and other modern materials, and forgets about alternative binders such as lime or earth. There is ample use of indigenous binders such as cowdung and some special earths, but the result is generally low quality. There is potential in Uganda for alternative binders, such as lime-pozzolanas which are already used on a small-scale, and it might also be possible to blend the traditional with the modern and come to improved indigenous binders.

With a per capita income of 60US$/year, and a bag of cement costing upwards of 6$ depending on the transport involved, the use of cement is also prohibitive in Tanzania; besides, it is not easily available in some outlying areas. Cement is nevertheless the ultimate dream of many dwellers, but a dream that often does not come true. It has also been noticed that the use of cement on earth walls often leads to poor results, with plasters coming off, where lime is much more appropriate. Away from the coast, though, the use of lime in housing is not very common. There is also a considerable use of indigenous binders, such as pozzolanas, earths, cowdung, but they are not very durable. The survey in Tanzania shows a noticeable difference in quality between houses where binders were used, and those that had not; the main reasons for not using binders where unaffordability (of the commercial binders), unavailability, and sometimes ignorance. The survey recognizes the need to introduce more appropriate binders, to train artisans accordingly, and to develop the necessaru codes and standards to allow their use in urban areas where building is more controlled.

Zanzibar and other islands such as Pemba and Lamu, as well as a few historical towns such as Mombasa and Bagamoyo, are a case apart. They have a long history of using limestone and lime mortars for walls, as well as plastering and rendering those walls with lime mortars, often including pozzolanic materials. Even to-date, there is a lot of lime production in those areas, and lime is still often used in housing. A lot of the traditional skills in using lime are getting very rare, but a number of conservation projects now allow lessons to be learned and skills to be preserved or developed again. The region as a whole can be made to benefit from this.

In summary, the following statements probably apply to the whole region:

- houses that do not use binders are generally of poor quality, showing amongst others erosion, cracking and attack by insects,
- cement is too expensive for most; a bag may cost as much as 40% of a monthly minimum wage in various areas, and cement plaster is 6-25 times as expensive as some indigenous plastering materials,
- lime is somewhat better priced, but apart from areas with a lime tradition such as Zanzibar, not regarded on par with cement and relatively unknown to the builders,
- overall, it is likely that no more than one tenth of low-income housing in the region has used cement or lime, largely due to their unaffordability or unavailability,
- there is a considerable use of indigenous binders such as special earths, cowdung, ash and pozzolanas with low to medium quality results; some of these options are under threat of disappearing resources or skills,
- with indigenous solutions that are not quite adequate, and commercial solutions that are generally too expensive, there is clearly a need to find intermediate solutions; the following points could be considered:
- blending commercial binders with low-cost or waste materials such as pozzolanas, in a proportion of up to 1:2, could reduce their cost, whilst maintaining adequate quality,
- the decentralized production of commercial binders could reduce their transport cost, which can be up to twice their production cost in some outlying areas, hence make them more affordable,
- the efficiency of lime production could be improved, reducing fuel costs and improving its quality, which could make lime more economical to use,
- research is required into the state-of-the-art of indigenous binder use, as a base for its improvement, e.g. by stabilization with lime or cement,
- alternative or innovative binders need promotion and demonstration,
- building artisans need training in the proper use of alternative binders as well as the most efficient and less wasteful use of binders in general.

Kenya

Daniel K. Irurah, T.J. Anyamba and Kigara Kamweru

Binders constitute an essential ingredient in building construction. Their appropriate use enhances durability, hygiene, aesthetic qualities and the strength of the structure. Conversely, inaccessibility or inappropriate use of the materials by owners and builders greatly impairs the quality of houses. There is an urgent need to evaluate the inaccessibility and inappropriate use of binders in low-income housing with a view of evolving appropriate intervention strategies. This study looks at binders in four categories:

1. High technology-based binders such as cement
Cement though highly centralized in Kenya in terms of production, is the most widespread binder in terms of availability. However its access is highly impaired by prohibitively high cost of the binder itself and both the complementary materials its use requires (such as aggregates) and its application technology (skills and tools).

2. Intermediate technology-based binders such as lime
Lime is much cheaper than cement. Its production is fairly centralized, and its complementary materials not very different from those for cement, but there is some difference in the application technology. Although relatively more affordable than cement, its cultural acceptability is very low mainly due to inadequate awareness of its performance potential and its application technology.

3. Indigenous technology-based binders such as cowdung and special earth
Cowdung and special earths were found to be more accessible in terms of availability and cost. However communities access to these binders is getting constrained by the forces of declining cultural acceptability and uncertain future supplies as a result of competition from alternative uses (for cowdung) and environmental degradation (for special earths)

4. Newly emerging binder alternatives, especially pozzolana-blended binders
Pozzolanas are natural or artificial materials which contain silica and/ or alumina; they are increasingly used in Kenya with cement clinker, cement or lime to make blended binders (see also Section 1).

When access to binders was evaluated in terms of cumulative cost per unit area of plaster (KSh/m^2) cowdung was found to be the most affordable at KSh5/m^2 followed by special earth (KSh20/m^2). Cement was the most inaccessible at KSh 123/m^2.

The key factors deemed to influence binder use are:

- availability and cost of binder materials within a given locality or community.
- availability and cost of complementary materials required.
- availability and cost of appropriate skills and technology for the application of the binders.
- level of cultural acceptability of the binder.

The proposed intervention strategies to enhance access to binders and their appropriate use fall into two broad categories: the development of alternative binders whose performance is culturally acceptable at a lower cumulative cost than cement or lime, and the dissemination of relevant application technologies for these alternatives.

Lime has a major potential compared to cement, cowdung or special earths. Although, like cement, its production is dependent on limestone deposits, lime technology allows for production on a much smaller and more decentralized scale, as is the case in Lamu and Wajir. The main alternative envisaged, is the production of blended cements or limes using non-industrial techniques. This can be done close to points of use and with various mix ratios which are more appropriate to the lower performance standards expected of binders in low-income housing construction unlike the stringent standards applied in the formal sector of the construction industry.

Research into the enhancement of the performance potential of the indigenous technology based binders through blending or stabilization with cement or lime would be a worthy starting point. Such research would also be geared towards enhancing the cultural acceptability of such binders and their future availability.

The training of artisans and owners of buildings is critical. In this respect, it should be recognized that binder-use is an art and a culture and hence the objective should be to inculcate an alternative binder use culture within the community. The aesthetic quality and creative involvement of owners and builders should be equally emphasized.

The rationale for the selection of case studies was mainly pegged to lime as a binder of special interest in the study. Consequently, seven case studies were selected. The scope of this study can be discussed in respect to three key factors:

Geographical spread: All the case studies were selected in such a way as to yield data which would allow the formulation of qualitative findings in relation to certain factors which were deemed to be most critical in influencing access to binders and the art of binder use. Because of this, the study has limitations in its geographical coverage, quantitative results and coverage of the whole variety of binder uses.

Socio-economic mix: The socio-economic groups covered by this study are mainly the owners and builders of low-income housing both in rural and urban areas. Indigenous skills and technology were also considered. The views of certain groups, such as renters or women, may have had proportionally less attention.

Binder varieties covered: The study covers both the conventional high technology-based binders like Portland cement and indigenous-technology based binders such as cowdung and ash. Newly emerging binder alternatives such as rice-husk and rice-straw ash which are used as pozzolanas are also covered. Lime is covered as an intermediate technology-based binder. Some problems were encountered during the fieldwork in comparing the quality of surfaces plastered with various binders and in establishing the relationship between this quality and their construction process, which presents a limitation of the study.

Variety of binders and their role in low-income housing construction

Building culture denotes the ways and means of shelter-provision within the community. The building culture can be analysed in terms of 'forces', 'processes', 'actors', and 'artefacts/products'. The 'forces' in binder-use culture would constitute the factors and needs which necessitate the use of binders in low income housing. Key among these factors are:

- *physical and technical forces:* The need to provide strength and weather/wear resistant surfaces so as to enhance the durability of the building.
- *cultural forces:* The need to constitute and express identity at various social scales (individual, family, clan, neighbourhood etc.) and express the cardinal values of the owner/builder and community, especially in relation to order and beauty.
- *socio-economic forces:* The need to invest resources towards enhancing the economic value of the building and the socio-economic status of the owner/users among the community.

The roles of the various actors in binder-use are often determined in terms of: gender; age; skills and market forces.

Binder categories applied in the study

Four categories of binders were determined and studied:

1. High technology-based binders, particularly cement, which is industrially processed on a large scale.
2. Intermediate technology-based binders, particularly lime, which is mostly also industrially processed, on a medium scale.
3. Indigenous technology-based binders, relating to traditional building methods and the long association people have had with them; they are usually produced or processed at a domestic scale, as and when needed, thereby minimizing procurement and transport cost.
4. Newly emerging binders, which in this case incorporate pozzolanas as additives to lime or cement; their use is relatively new in Kenya. The more common pozzolanas include diatomite, rice husk ash, and volcanic ash or tuff, all of which have been used in Kenya. So far, they have been mainly industrially processed, on a medium scale.

Contrary to the indigenous binders, the use of industrially processed binders has a formality about it such that an accepted and inflexible way to use the product soon emerges and is generally understood. Industrial binder products require trained or skilled artisans for their application. In this process rules of thumb and fixed processes tend to dominate the binder-use culture such that the end product becomes easily standardized and predictable.

Table 48: Comparison of key binder characteristics

Industrial technology binders	Intermediate technology binders	Indigenous technology binders
Cement	Lime	Cowdung, ash, special earths
mass production	local level production	domestic scale production
geared to formal sector	may be designed to suit different sectors	Used in traditional buildings
Expensive Usually involves specialized personnel	Moderate pricing Suitable for artisanal production	Fairly cheap Usage well understood within respective communities
Full utilization of potential pending	Culture of application still developing	High level of art and cultural expression
Relatively new in rural communities	Used in some areas traditionally. May need to be introduced in others	Used in traditional buildings

Production and distribution of binders

Despite rail haulage being far cheaper, road transport is the major means of transporting cement or lime to most destinations in many parts of Kenya's hinterland. Freight accounts for approximately 25% of the delivered cement or lime. In as far as transportation costs go, rail tariffs range from KSh 1.0 to KSh 1.5 per tonne/km (decreasing with increasing distance). Road transport tariffs are much higher, ranging between KSh 1.20 and KSh 3.20 per tonne/km. For shorter distances, the unit rate goes up, while for longer distances lower rates are negotiated; average rates are about KSh 2.20 per tonne/km.

According to the current logic of cement prices, the more distant customers are being subsidized by those located close to the cement plants. The following are some price quotations provided by the East African Portland Cement Company (EAPCC).

Table 49: Price of bagged cement per tonne (1990-94)

Year	Price per tonne of bagged cement	
	KSh	US$
January 1990	2,383	109.7
May 1991	2,750.35	99.22
May 1992	3,047.35	96.28
September 1992	3,496	104.9
June 1994	7,846	142.65
November 1994	7,508	187.7

Source: East African Portland Cement Company

As can be seen from the above figures, the cement price in Kenya Shillings has more than trebled between 1991 and 1994. However based on the dollar, cement prices reduced between 1990 and mid-1992, to nearly double again between September 1992 and June 1994. Since income levels in Kenya have not trebled in the last four years, cement is unaffordable to most low-income earners.

High technology binders

Currently the two cement producing plants produce mainly Portland Pozzolana cement. Ordinary Portland cement is produced on order. By adding between 15-30% of pozzolanas to the clinker the plants are able to increase their production capacity. It would be expected that Portland Pozzolana cement would cost less than ordinary Portland cement, but this is not the case.

Most of the cement produced in the country is used in the formal sector of the construction industry. It is not easy to demarcate the formal and the informal sector of the industry in Kenya, as statistics concerning buildings cannot be easily obtained, many buildings are built without seeking local authority approvals and also the fact that no surveys have been carried out in rural areas to ascertain the level of building construction. A large proportion of all private construction belongs to the informal sector, of which probably less than 10% use cement or lime as binders. The majority of the informal sector households are constituted by semi-skilled and unskilled workers. Their average monthly wage is approximately KSh 1,500 ($27). This is equivalent to the cost of three bags of cement, which makes these binders ill affordable to this category of workers.

Intermediate technology binders

The two lime producing plants in Kenya have traditionally produced lime for the industrial and agricultural sector. It is only recently that lime is increasingly getting used in the construction industry. Most consumers therefore still purchase their lime directly from the factories at Koru and Tiwi.

In addition to the two lime producing plants in the country, there exists small-scale lime production in Lamu and Wajir based on indigenous technology.

The current cost of lime ranges between KSh6,000 ($109) per tonne for the grey lime of Koru and KSh 9,000 ($170) per tonne for the white lime of Tiwi. However it should be noted that one tonne of lime is equivalent to two tonnes of cement by volume. This makes lime much cheaper than cement, in its day-to-day use on the building site, where mixes by volume are commonplace. However, to make up for its lower quality, lime mortars should be somewhat richer in volume of binders than cement mortars, which again reduces their advantage.

Indigenous technology based binders

Hard varieties of laterite are used for foundations of buildings in dry regions. Termite hills, which are known to resist rain, can be pulverized and used as a stabilizer for sandy soils. Special earths do exist all over the country, however their continued use leads to environmental degradation.

Cowdung is generally used in its natural state or as a stabilizer for clay or sand. It is used as a floor and wall finish in most houses constructed with mud. Cowdung as a stabilizer is valued mainly for its reinforcing effect (on account of the fibrous particles) and ability to repel insects. Water resistance is not significantly improved, while compressive strengths are reduced.

There however exists competition for the use of cowdung. Cowdung produces excellent manure for organic farming, which renders it valuable as an agricultural input. Therefore, one may have to look for alternative stabilizers of soil, such as lime.

Newly emerging binders

Pozzolanas are natural or artificial materials which contain a substantial amount of silica. Some essential small quantities of alumina may also be present. They are not cementitious themselves, but when finely ground and mixed with lime, the mixture will set and harden at ordinary temperatures in the presence of water, like cement.

Pozzolanas can replace 15 to 40% of Portland

cement without significantly reducing the long term strength of concrete. Pozzolanas contribute to cost and energy savings, help to reduce environmental pollution and, in most cases, improve the quality of the end product. Natural pozzolanas are essentially volcanic ash from geologically-recent volcanic activity, and a few other materials such as diatomites. Artificial pozzolanas result from various industrial and agricultural processes, usually as by-products. The most important artificial pozzolanas are: burnt clay, pulverized fuel ash, ground granulated blast furnace slag, rice husk ash and bagasse ash.

Gypsum is readily available either as natural gypsum or as an industrial by-product, and is cheaper than lime or cement (produced with less energy and equipment). Gypsum when mixed with water hardens rapidly. Adobe blocks stabilized with gypsum require lengthy curing, but can be used for wall construction. Gypsum when used as a stabilizer produces blocks with low shrinkage, smooth appearance and high mechanical strength. In addition, gypsum binds well with fibre(particularly sisal), is highly fire resistant and is not attacked by insects and rodents.

The use of rice husk ash as an alternative binder in Kenya has not been fully explored yet. One of the other possible by products of rice which could be used for ash is the straw, which represents a considerably larger quantity (per tonne of paddy) as compared to the husk. However the use of straw ash as a binder gets competition from other uses e.g. production of straw board, paper and animal fodder.

Masonry cement as manufactured in Kenya is essentially a blended cement. The materials used for blending are a mixture of diatomite, gypsum, limestone and kaolin. (The Kenya Bureau of Standards is in the process of producing a standard for this product.)

Key observations on case studies
Kinyago village - Nairobi
Kinyago village in Nairobi was selected to represent informal housing neighbourhoods in major Kenyan cities. It is a newly-built settlement which owes a lot to the intervention of an NGO, the Undugu Society of Kenya. Undugu organized community self-help efforts by assisting in the layout of the residential units on site, and providing 4 bags of cement and 12 wheelbarrows of sand per individual unit; owners were to provide for labour and other materials.

Having provided the enabling environment, Undugu expected the owners to put a constant effort in upgrading their dwellings on their own. This has taken place only to a very limited extent. The case study is therefore relevant in two ways: it shows the potential of binder use in the informal housing sector for low-income groups in urban areas, as well as the advantages and disadvantages of direct provision of materials and services to the owners as an intervention strategy.

The neighbourhood has no locally procured binders of any kind. Cement is the most prevalent binder in terms of availability and use. An isolated case of special earth wall plaster and floor was observed. Availability and use of other binder alternatives like lime or the newly-emerging binders are almost nil within the neighbourhood or in nearby hardware shops. However, large hardware stockists within the city do sell lime at a cost of about KSh 150 per 25 kg bag ($2.7); this would involve additional transportation cost of between KSh10 to 20 per bag. For cement, the procurement cost was around KSh 410 per 50 kg bag ($7.5). Transportation to the site is mainly manual and relies on wheelbarrows. An additional transportation cost of KSh 10 per bag is often charged. The non-use of lime in the neighbourhood can therefore be attributed more to a lack of awareness of the potential of the binder and the fact that owners and builders view the material as inferior to cement, than to its price.

Responses from community and opinion leaders, as well as owners of houses where cement had been used showed that enhancement on durability and the aesthetics of resultant surfaces were the key factors which influenced the choice of the binder. Reduction on maintenance requirements and ease of maintaining clean surfaces were some of the other factors cited.

The major constraint was the cost of procuring the binder, other factors like distance to procurement point, consistence in availability, quantities in

Table 50: Quality rating of surfaces in relation to binders used

Binder	Houses surveyed	very high	high	medium	poor	unavailable
Cement	78 (100%)	7 (9%)	29 (37%)	26 (33%)	13 (16%)	4 (5%)
Cement/lime	27 (100%)	3 (9%)	8 (25%)	17 (53%)	1 (3%)	3 (9%)
Lime	3 (100%)	-	-	-	2 (67%)	1 (33%)
Special earth/ lime	3 (100%)	-	3 (100%)	-	-	-
Special earth/ cement	4 (100%)	-	1 (25%)	3 (75%)	-	-
Cowdung	12 (100%)	-	-	4 (33%)	7 (58%)	1 (8%)

packaging, and the cost of labour for qualified builders or artisans were secondary. The residents did not know of any other binder alternatives.

Suggested intervention strategies for Kinyago include:

- subsidies from community based organizations; the case of Undugu Society's assistance was frequently mentioned.
- provision of funding through self-help or co-operative loans.
- enhancement of income generation of households.
- strengthening of community self-help groups.
- dissemination of information on availability and use of cheaper alternatives.

The quality of wall and floor surfaces in 20 out of the 22 (91%) houses surveyed ranged from poor to very high, depending on the maintenance level. Amongst the 8 houses surveyed where no binders had been used, six were constructed of mud infill on wattle poles and railings. The resultant quality of the wall surfaces was very poor in almost all of these houses.

Gititu Sublocation - Nyeri

Gititu sublocation is a rural centre located in Nyeri District of Central Province. The case study was selected to represent a rural setting far from a lime or cement producing plant, but within areas which enjoy good transportation and binder distribution networks in terms of roads and outlets for building materials.

Cowdung and cement were the two prevalent binder alternatives available in the neighbourhood. An isolated case of lime use was observed in one homestead, but it was evident that the local hardware shops at Muthinga Market do not stock lime at all. Although cowdung has been available free of any financial cost for a long time, there is growing pressure to commercialize it as a result of increasing demand and the inadequate production rate. Currently, cowdung is occasionally sold at about KSh10 per container of about $0.05m^3$. The shortage of the material is also related to the competing use of cowdung as manure which is in great demand for use in agriculture. It is therefore evident that future availability of cowdung as a binder is threatened. Besides being the most widely available binder in Gititu, cowdung was also the most commonly used binder in the neighbourhood. This is because of its general low cost, its availability, aesthetics and durability. However, cowdung had two major drawbacks which discourages its use: it is of lower quality when compared to industrially processed binders like cement, and its cultural acceptability is diminishing as it is associated with the negative socio-cultural values of poverty, low-social status and backwardness.

Cement was the second most prevalent binder in terms of availability and use. It can be locally procured in small quantities of one or two bags at a time from a hardware shop in Muthinga Market. Its cost there was KSh 430/50 kg bag ($7.8), and transport to homesteads is mainly manual (in wheelbarrows or on human backs or shoulders). As the second preferred binder, cement was mainly chosen by owners and builders for its durability and minimal requirement for regular maintenance, its enhancement of the aesthetics, and its high level of cultural acceptability. Cement had one major drawback: its high cost which severely limited its access. Availability and the cost of hiring skilled artisans for the application of cement-sand plasters was also cited as a major constraint. The cost of an artisan, or 'fundi' was about KSh 150/day.

Owners and builders of houses, were unaware of other binder alternatives than cowdung and cement. There was no case where the newly-emerging binders had been used. Such materials were also not available in the hardware shops in Nyeri town, the provincial headquarters of Central Province.

The main intervention strategies which were suggested to alleviate the constraints cited are:

- introduction and popularization of cheaper but high quality binder alternatives
- training of builders and artisans in the application of a wider variety of binders.
- improvement on the quality of cowdung to enhance its cultural acceptability.
- government subsidy on the price of cement for the construction of low-income housing.
- control of the procurement cost of cement.
- improvement of income generation to enhance affordability.
- creation of funding opportunities (loans, fund-raising, etc.).

Five out of the 6 houses surveyed where cowdung had been used as a binder were rated to be of medium quality in terms of wall surfaces. Amongst the 8 houses surveyed where cement had been used, 2 had wall surfaces rated to be of very high quality, 2 of high quality and 4 of at medium quality. The use of paints on walls and slurry with iron oxide on floor screeds was commonly observed. Due to the wisespread use of cowdung or cement as binders, no houses were surveyed in Gititu where no binders had been used.

Majengo - Nyeri

Majengo is one of the oldest low-income housing neighbourhoods in Nyeri town. The original settlement was constructed of mud and wattle, plastered with lime and roofed with tin sheets. Very limited construction and rehabilitation has taken place since the initial construction. Majengo-Nyeri was chosen as a case study because of its proximity to Gititu, and also because it had a construction technology very close to that used in Gititu sublocation. But unlike Gititu, Majengo-Nyeri had several houses where no binders had been used which could be surveyed for comparative analysis.

Four binder alternatives were available and had been used in Majengo-Nyeri. These are cement, lime, special earth, and cowdung. Lime and cowdung are materials which had been used in the initial construction of the settlement, but their use seems to have been discontinued. Cement and special earth seem to be the only materials which are prevalently available and in use today, although the dwellers are aware of lime and cowdung as potential alternatives, from their use in the past. Cement is regularly available from hardware shops in Nyeri at a cost of KSh 430/50 kg bag ($7.8), while special earth is around KSh 20/container.

Special earth is the most commonly used binder due to its easy availability and affordability, as well as the enhancement of durability and aesthetics of the wall surfaces. Cement was chosen for its superior performance in enhancement of durability, and aesthetics of the wall and floor surfaces and also because it minimizes on maintenance requirements. One respondent also gave 'enhancement of investment value' (cement plaster attracts higher rents) for rental housing as a key factor which influenced the choice of cement.

The key constraints in using these binders, and a reason for not using binders at all, were the high cost in the case of cement and the poor performance in the case of the special earth *ngumba*. Responses on possible intervention strategies focused on the cost of procurement of cement as the most desirable binder. The suggestions include:

- Government to subsidize cost of cement.
- Government to build houses for the people at subsidized cost.
- improvement of the performance of the special earth in combination with lime, while keeping its cost affordable to a majority of the owners and builders.

Depending on the binder used and the maintenance by the different owners, the rating of the houses surveyed ranged from high to poor. In cases were binders had not been used major design and structural defects were detected, which included lack of differentiation in space, uneven floor surfaces and poor lighting levels, leaning walls and poles eaten by ants or rotten. Maintenance defects noted were: dusty floors, mud infill falling off, mud infill cracking with gaping holes and damp floor surfaces.

Passenga Settlement Scheme
Passenga is situated about 15km from Ol Kalou town in Nyandarua District. Passenga Settlement case study was selected to represent rural neighbourhoods far from a cement or lime production plant, and within areas where the transportation and binder distribution networks in terms of roads and outlets for building materials are scanty or in very poor conditions.

Cowdung and cement were the only two prevalent binder alternatives available in the neighbourhood. Cowdung is procured almost free of cost just like in Gititu. However, the material is less popular in this neighbourhood. It was evident that most of the cowdung plastered houses were built during the early days of the settlement. Most of the more recently constructed houses were either built of stone masonry with cement-sand plaster, or non-binder based construction technologies, especially using timber off-cuts.

The procurement cost of cement at Ol Kalou or Nyahururu was about KSh 500/50 kg bag ($9). An additional cost of about KSh 35/bag is charged by the few public service vehicles *(matatus)* up to Passenga shopping centre. From here, the cement is transported by donkey carts or manually to the homesteads. Distances involved are often long due to the large tracts of land per homestead.

Cowdung was the most widely available and also the most commonly used binder in Passenga Settlement. Factors which influence the choice of this material are: low cost, availability, aesthetics, durability and thermal performance. As observed earlier, most houses where cowdung had been used were built about 30 years ago and they are in a state of disrepair. Where attempts have been made at replastering, raw cowdung without ash had been used and the resultant surfaces are of very poor quality compared to the older surfaces or to similar surfaces in Gititu sublocation. One of the reasons cited for this problem was that there has been an attrition in cowdung application skills, especially because the practice of communal cowdung plastering of houses is non-existent here, unlike in Gititu sublocation where it is still a common phenomenon. People in Passenga Settlement seem to be also intensively engaged in other economic activities like farming and they hardly get time to maintain their houses. Consequently some of them have switched to masonry construction, a practice where cement is the main binder or to binderless construction technologies, especially the one using timber-off cuts on a timber framework. Another major factor contributing to the decline of cowdung use in this neighbourhood is that the material is regarded as a material for the poor.

Cement is the second most prevalent binder in terms of availability and use. The material was mainly chosen for its superior performance. Responses concerning factors which influenced owners of houses to choose the binder include: enhanced durability, minimal maintenance, aesthetics and status.

Key constraints in using binders include: the cost of cement and its long supply distance, as well as the availability and cost of qualified builders. Owners and builders of houses were unaware of other binder alternatives besides cowdung and cement.

Intervention strategies suggested in Passenga include:

- Government control of the price of cement.
- improved road linkage between Passenga Shopping Centre and Ol Kalou.
- improvement of public service vehicles between the major towns close by (Ol Kalou, Nyahururu, Nakuru) and Passenga Shopping Centre.
- introduction of alternative building technologies, for example the use of mud mortars pointed with a cement mortar as used in construction of colonial settlers houses which have lasted up to the present.

All of the houses surveyed where cowdung had been used as a binder were rated to be of 'poor' quality in their wall surfaces. The 13 cement plastered houses surveyed had their wall and floor surfaces rated as 'high' (38%) and 'medium' (46%) quality.

Where no binders had been used, the timber off-cut construction technology was emerging to be the most prevalent type amongst the newly constructed houses. As anticipated at the conceptual stages of the research work, the timber off-cut practice of house construction had evolved as a substitution strategy for the use of binders given their constraints, related to availability, cost, and performance.

Majengo and Misakwani villages, Machakos

Majengo village is a low-income housing neighbourhood situated within Machakos town. Like most other 'Majengos' in Kenya, it is predominantly a muslim village. The main criteria for selecting Majengo and Misakwani villages in Machakos was to collect data which would give some insight into the culture of special earth application as a binder. The choice of nearby Misakwani village was made so as to carry out the survey on more recent construction which was not possible in Majengo Village itself.

Three binder alternatives are available and used in Majengo and Misakwani. These are cement, lime, and special earths. However, special earth is the most commonly used binder alternative, and cement is the second most prevalent binder used. It is apparent that lime had only been used in the past (during the 1950s) among the houses surveyed. Cement is available from the hardware shops in Machakos town. It costs KSh 430/50 kg bag ($7.8). Transportation to the sites was mainly manual. Special earths used are procured form the Iveti Hills, at a distance of about 6 km from Machakos town. Two major special-earth types have been used. These are: Ntha (Red or Pink coloured special earth) and Mbuu (White/yellow coloured special earth).

As the most commonly used binder in Majengo and Misakwani Villages, special earth is preferred for its ease of procurement in terms of availability and cost, as well as for its enhancement of durability and aesthetics of the wall surfaces. Responses from community leaders as well as the three owners of surveyed houses where cement has been used show that enhancement of durability and the aesthetics of resultant surfaces are the key factors which influence the choice of cement as a binder.

Responses from community leaders and owners of surveyed houses where no binders have been used in this neighbourhood, indicate that the cost of procurement in relation to cement, and distance to procurement points and the lower quality of binder in relation to special earths constitute the major constraints which inhibit use of binders in the neighbourhood.

Besides cement and special earth, there seems to be only a limited awareness of the availability and performance of lime as an additive to cement. Although lime has been used in the older buildings, the use of lime in isolation is not evident in the newly constructed houses.

The binder situation in Machakos could be improved by:

- price control of cement by the Government.
- a community-based initiative for demonstration houses using indigenous and locally available materials so as to train the younger generation before the skills die.
- alternative funding for house construction (e.g. formation of co-operatives or other self-help groups)
- introduction of alternative binders of high quality but cheaper.

The quality of wall and floor surfaces where binders had been used was high to very high in 21% of all cases, medium in 57% of all cases, and poor in 7% of all cases, with the remainder being unrecorded. Apart from design defects, houses where no binders had been used have problems of leaning walls, timber poles exposed to weather and insect attacks, erosion of mud mortar, and maintenance related defects of wall and floor surfaces.

Koru and Muhoroni locations

Koru and Muhoroni are rural neighbourhoods in Nyanza Province. The residents are mainly people who settled there after independence and most of the houses were built after 1964. The case study was selected to represent rural neighbourhoods close to an industrial lime producing plant. The data collected here would help to evaluate if closeness to such a plant has any significant impact on binder use in such neighbourhoods. The area enjoys good quality transportation and distribution networks in terms of roads and outlets for construction materials.

The Koru/Muhoroni case study has two binders which are abundantly available. These are cement and lime. However, cement is the most commonly used binder, and lime is only used as an additive to cement. Cement was readily available from hardware shops at Koru and Muhoroni shopping centres at a cost of about KSh 500/50 kg bag($9). The transportation is mainly by public service vehicles at a cost about KSh10/bag. Lime is also readily available from the nearby hardware shops. The procurement cost is KSh 150/25 kg bag($2.7), and transportation cost is the same as for cement. Lime is commonly used as a complementary material to cement. Cowdung is also available but is not prevalently used.

The choice of cement is influenced by:
* predominance of masonry construction using quarry stone and murram blocks.
* easy availability of binders from the close by hardware stockists in Koru and Muhoroni shopping centres.
* good road network.
* high income levels for most households especially from sugar-cane production.
* availability of skilled artisans most of whom are trained at village polytechnics.

In the rating of the main factors which inhibit binder use in Koru/Muhoroni, opinion and community leaders gave the cost of cement as one of the major constraints. Two out of the four leaders interviewed said procurement price of cement was severely inhibiting the use of binders. Most of the other factors were of no major effect in inhibiting the use of binders. The major constraint cited in relation to lime was the unavailability of qualified builders and artisans who have relevant skills in lime application. Lack of awareness of alternative binders was also cited as a major constraint by owners of houses where no binders had been used.

Owners and builders of houses in this neighbourhood are aware of the availability of lime and its performance when used as a complement to cement. It was evident however that the use of lime as a binder in itself is very rare. Some of the reasons given for this include:
* lime is a weak binder and it is unlikely to perform as well as cement.
* lime takes very long to set and cure.
* there are no artisans who have the necessary skills of using lime on its own.

It is also evident that cowdung was available in the neighbourhood, but has been used in very few houses. No other binder alternatives were observed.

Suggested intervention strategies include:
* introducing alternative binders and boosting awareness through
  a. training of builders in the use of the alternative binders.
  b. putting up demonstration units using alternative binders
* general improvement of living standards, especially through enhancement of income-generating activities.

The quality of the walls ranged from fair to high, where cement and lime had been used. On the other hand ten houses with no binders used were surveyed. Three of those were of posts and compressed murram construction. The structure and surfaces (floor and walls) were rated to be of very high quality even though no binders had been used. This might be attributed to good workmanship and the fact that all three houses had been built in or after 1984, which is fairly recent. The other 7 houses had serious design, structural and maintenance defects.

Homa lime factory - staff housing, Nyanza

The case study is located in Koru, Nyanza Province. The houses surveyed are mainly the junior staff housing for the lime factory workers. Although the houses are employer-built housing (which is essentially formal) the case study was selected to help evaluate the quality of wall and floor surfaces with the use of lime as a binder.

One of the key observations in this case study is that the quality of wall and floor surfaces was generally high. However, it was evident that inspite of the high availability of lime from the factory - hence the relatively lower, cost the company had not used lime as a binder on its own. The over dependence on cement as the main binder was rather surprising in this case study given the fact that the company is fully aware of the unexploited performance and potential of lime as a binder. This lack of confidence in lime had spread to the surrounding rural neighbourhoods where inspite of the widespread use of the binder, no case was observed where lime had been used as the only binder, it was always used as an additive to cement.

Quality of wall and floor surfaces

Quality rating of surfaces in relation to base material of walls: 44% of the masonry houses surveyed had their surfaces rated as high or very high quality while only 3% of the houses had their surfaces rated as poor. This was a very high performance when compared to mud and wattle as a base material where only 22% of such houses were rated as high or very high quality. Almost three quarters of the mud and wattle houses were rated as medium (40%) and poor quality (34%).

Quality rating of surfaces in relation to age of house: Most of the houses surveyed are 20 years old or less. The high performance of cement and cement/lime even in relation to age of houses is evident. This pattern strongly supports the owners' and builders' view that cement and cement/lime do enhance durability and the aesthetics of surfaces better than cowdung or special earth.

Quality rating of surfaces in relation to experience of artisan: The experience of artisans was rated in terms of houses built rather than the number of years in the trade. Since most of the cowdung and special earth houses were owner-built, it was not possible to use this parameter on such houses. Elsewhere, there does not seem to be a clear relationship between experience of the artisan (fundi) and quality of the surfaces. It may be more a question of skills but without commitment to certain quality standards, this will not necessarily yield better quality surfaces. While the need to save costs could be cited as a major factor in the pattern observed, field observations indicate that most fundis were not too sure how quality is achieved for walls and floors.

This was different in the case of cowdung and special earth plasterers. Most of the owners who did this work had a sense of pride in their work and it was clear that they were striving to realize the best possible quality of surfaces with the material.

Quality rating of surfaces in relation to mix ratios and techniques of application: It was evident that the variations encountered in the case studies did not have major implications on the quality of surfaces. Various mix ratios were cited (see also furtheron), but it was evident that these were often disregarded depending on the ability of the owner to acquire the required amounts. It was therefore clear that the feel of the fundi as to the limit of possible ratios was more critical than what cited was in practice.

Observations in the field showed that cement and cement/lime had a more standardized practice of mix ratios and application techniques in terms of number of layers, thickness of layers and curing period while cowdung and special earths depended more on the feel of the builder. Most of the curing requirements for cement and cement/lime were the responsibility of owner: spraying with water, avoiding dry curing or exposure to sun, allowing for the appropriate period, etc. The same applied to cowdung and special earth.

Art and culture of binder use in the case studies

Kenyan societies have a well developed traditional building culture as evidenced by the architecture of the various communities. This architecture is characterized by several parameters:

- the use of locally available materials.
- building to directly address the issues of climate and the environment.
- the use of communal labour.
- an attitude towards decorative elements and patterns.

With time, this set tradition has been disturbed by the introduction of new materials and new forms:

- the desire for durability and permanence
- the depletion of some traditional material and consequent need to adopt and use new materials.
- the need for new status symbols.
- new building practices result from this evolving set of values.

New characteristics of the building culture assert themselves with new forces and distinguishing aspects such as:

- new industrially processed materials like cement have been introduced
- with these new materials, new standards, and disciplines are put into place.
- also introduced was the element of cost as a major determinant of what should be built or not.
- new forms also relate to status.

The cases studied for this report had these characteristics to different degrees depending on the traditional way of building and the extent to which new attitudes have been involved. In the use of binders, it was clear that some were more popular than others in the different localities as follows:

Table 51: Schedule of commonly used binders in the case studies

Case study	Commonly used binder
Kinyago village, Nairobi	Cement
Gititu Sublocation, Nyeri	Cowdung/ash, cement
Majengo village, Nyeri	Special earths
Passenga Settlement, Ol Kalou	Cowdung, cement
Majengo/Misakwani villages, Machakos	Special earth
Koru/ Muhoroni locations, Kisumu	Cement/lime
Homa Lime factory	Cement/lime

In each area different forces tend to manifest themselves and each material thereby develops a culture of application and utilization. Thus the examination of the art and culture of binder use can be explored through the manifestations of the more commonly used materials.

The culture of application of cement

At the local level cement is normally sourced

through small hardware shops or building supplies stores located at the nearest towns or market centres.

Cement is a widely known binder whose qualities and general performance provide a datum for the assessment of other binders among people familiar with it. Cement is used as a plaster or as a mortar because it is highly durable and provides strength beyond most requirements.

Table 52: Use of cement in case studies surveyed (prices: June 1994)

Case study	Total using binders	% using cement	Cost/50 kg bag KSh	$
Passenga	20	55	480	8.7
Kinyago	22	77.3	410	7.7
Gititu	14	50	430	7.8
Majengo-Nyeri	9	22.4	430	7.8
Misakwani	13	23.1	430	7.8
Majengo-Machakos	9	37.5	430	7.8
Koru/Muhoroni	50	100	500	9.0
Homa Lime factory	19	100	500	9.0

Cement is now expensive to buy and use, and this factor has led to the identification of this binder as a material for the rich. The cost of cement in Kenya today (mid-1994) ranges from around KSh 420 ($7.75) in easily accessible locations like Nairobi to KSh 500 ($9) in Kisumu. This would be equivalent to a ten day wage for an unskilled construction worker. A status symbol has therefore evolved around the use of cement which is complemented by its general formal aesthetics.

The adoption of cement to a lower technology is an important pointer to the heralded position of communally based technology. Though cement builds to the language of mud, the introduction of the superior material does not automatically lead to the adoption of its definitive technology.

The process of using cement in low-income housing construction relies on a developing tradition commonly held by a cadre of craftsmen. This knowledge differs somewhat to the recommendations available to practitioners of higher technology and demonstrates a degree of innovation and accommodation towards easier and cheaper practice. The recommended use of cement as a binder will involve the mixing with several complements notably sand and water. The quality of the cement is commonly referred to in mix ratios that represent volume units.

In almost all cases where builders were interviewed, the units of measure were the implements of the work. It was noted that a bag of cement was usually related to a wheelbarrow of aggregate (e.g. sand) although recommended practice requires equivalent volume units. This cost saving measure produces a mix that would be regarded as inferior in the formal sector of the construction industry. While this may be regarded as producing inferior quality, it is worth noting that the recommendations relate to the very highest qualities of products. No standards have been set for low-cost construction and these practices may be setting the matter right by applying the proper standards for this level.

Another innovation geared toward cost cutting was the substitution of sand with quarry dust. This measure, while saving on costs, reduces the achievable quality of cement, whose performance is predicted on the percentage of air voids in the mix. This however does not seem to be a widely appreciated fact.

A major complement of cement is water. Whereas high technology usage of cement requires a measured use of water emphasizing the minimum amount required to mix, there is a liberal attitude towards the use of mix water in informal housing.

The mixing usually happens on the ground with no special place being prepared. However care is taken to minimize the presence of dirt and organic material in the final mix.

Builders who responded to the survey, seemed to vary within a narrow band in their use of mix ratios. There was also inconsistency on the use of batching measures with cement being batched differently from other materials. The commonly used mix ratio was one part of cement to 4 parts of sand or quarry dust (1:4) for plaster on the walls while for floors it was 1:3.

There is a well developed culture of tool-use and most artisans have the correct tools to handle cement plaster.

Use of cement invariably involves specialized personnel who have had formal training or have been apprenticed for some time with more experienced practitioners. Most artisans tend to start as young men who may have dropped out of formal school. However the study revealed a mean of 20 years experience in construction which would indicate a predominance of middle aged men. There is a predominance of apprenticed artisans. Formally trained craftsmen represented only 27% of the total; Koru had the highest number.

The owners role in decision making is crucial at the stage of deciding whether or not to use cement. The role here is a decisive one as the allocation and appropriation of resources is decided.

The culture of application of cowdung/ash

The use of the cowdung/ash combination as a binder is a widespread practice in the rural areas of Kenya and is one that can be traced to pre-colonial days. This long history is one reason why cowdung and ash as an additive has remained popular in certain rural communities.

Construction using cowdung is mainly a communal activity. This process assumes a knowledge of the material, and tasks are not assigned on the basis of specialized knowledge.

Cowdung is obtained when it is still fresh and poured into a shallow pit dug for the purpose. Ash and the appropriate amount of water are added and thoroughly mixed. The point of satisfaction is determined by the colour - a grey colour is preferred. Women mainly use their feet and hands to mix the ingredients. The composition of this binder is based on the one major resource: cowdung itself, to which various ingredients can be added to improve its qualities. The more common additive is ash, which is said to imrpove the cracking resistance and the colour. Ash is also said to improve the water and pest resistance of the cowdung although this could not be well established.

In Passenga Settlement, cowdung was mixed with ash in varied ratios. It seems the ash is seen to perform satisfactorily if it is about 1/4 by volume of the cowdung. But situations were encountered where the amounts were equal. The variation seemed to be caused by the different perceptions of the builders as to the role of the ash. In Gititu sublocation sand was also an additive. In this case there was a different attitude towards ash and a higher proportion than the Passenga case was used. Only two of the six houses surveyed used cowdung and ash with no additives. Three added sand to the cowdung/ash and one used cowdung with lime.

It is interesting to note that though the work is fairly taxing, this process is usually undertaken enthusiastically and does not have a stigma attached to it. No evidence has been found of specialist cowdung applicators. It would seem safe to state that they do not exist and that the knowledge of how to use cowdung is widespread in the respective communities. The use of communal labour in the application of cowdung/ash as a binder has been identified as a common phenomenon. It is however not universal and individuals can and still do undertake the task with family labour. The decision to use or not to use this particular material is however taken by the owner without the need for outside help. Responsibility of the upkeep of surfaces so treated reverts back to the owners immediately after. This is normally a woman's domain which reinforces the traditional view of women as custodians of shelter.

The general result of cowdung-use as a binder is an informal, low technology and low cost building. What is a characteristic feature in this binder use is the freedom exhibited by its users in the expression of the art of building. The main weakness with cowdung plaster binder has to do with its durability in the weather conditions.

The culture of application of special earths

Special earths were found to be predominantly used in the Majengo /Misakwani villages in Machakos. The option to use special earths seems to be propelled by the availability and therefore economic nature of the material. With a wide base of knowledge, it is also a material that is comfortable to deal with, its procedures being properly understood.

Special earths are prepared in a rather simple process which does not involve any particular skills. The essence of the process is the creation of a slurry by mixing the earths with water. The unit of measure is the tin or debe ($0.05m^3$) a half of which is mixed with about 40 litres of water. The slurry is mixed to a viscosity similar to emulsion paint with the coarser aggregate removed through sieving. Besides water, special earths tend to perform without other additions. However in a number of cases an oil skirting was found to be applied on the finished wall for water proofing and protection against ants. Another treatment was where the earth was mixed with an amount of salt and a small packet of blue detergent and the resultant slurry boiled. Respondents indicated that this procedure improved the adherence to walls. The process of using special earths is beginning to have a cost element both in terms of money and the environment.

Binder-use has developed a hierarchy of use where the lower the technical expertise required, the more widespread these skills are in a given area. Special earths do not require sophisticated knowledge and most people will be familiar with the process. However cultural issues may mitigate as to how the labour will be divided.

There is widespread satisfaction with the performance of special earths as a binder. Special earths are predominantly used on walls and the expression of satisfaction came from 86% of respondents. The overall picture that results from use of these earths is one of aesthetic preoccupation. Special earths seem to give an opportunity to use colour, moulding, and decorative patterns. These opportunities are heartily taken and their is a splash of colour in the areas using these materials.

However the sense of freedom does not extend to the basic forms used. Most of the houses using special earths as a binder have used burnt brick as the basic walling material (70%). The bricks being of regular shape and standard dimension, instill a discipline in the way they are laid which governs the final regularized and straightened form of house that is prevalent in the areas surveyed.

The culture of application of lime and cement/ lime mix

A widespread use of lime was encountered in the Koru/Muhoroni region of Western Kenya. This is an area with a limestone processing plant and the easy availability of lime is the primary reason for its widespread use. The use of lime here goes hand in hand with limestone blocks. One would assume that acceptance of limestone as the base material would lead to a choice of lime as the natural binder.

Lime has well known cementitious qualities and people living close to lime quarries know this. Lime is used as a plaster and as a whitewash for buildings. The wider use of lime is greatly hampered by the lack of full appreciation of its potential. A distrust of lime is detectable among the communities using the material. Thus despite despite widespread use in Koru and Muhoroni, a reluctance still persists regarding the use of lime alone. Cement is normally brought in to complement the lime. The two companies that process lime industrially have attempted some marketing and popularization of lime but this information has not reached the point of making a significant difference to the awareness level of lime as an alternative binder material.

The use of lime observed in the survey was in mixture with cement. In Gititu, Nyeri where some awareness of lime was encountered, the reasons given for the lack of wider application was the need to have protective gloves in a culture where people are used to handling material (mainly mud and cowdung) with bare hands. There was a general lack of skills in the application of lime and apprehension to the more strict precautions that use of lime demands. Lime as an intermediate technology based material may require some transportation from its point of source. In areas where other binders were prevalent, it is clear that knowledge of lime is limited and no skills have developed on its use.

In the areas surveyed, lime is sourced as a processed product virtually ready for use. Lime was used in mixture with cement as a mortar and as a plaster. When mixed with cement, the ratio applied is usually a 1:1 where the lime replaces an equal amount of cement by volume. The two materials are mixed together and used like cement. Another common application of lime is in whitewashing walls, which is seen as increasing the aesthetics. There are quite a lot of undeveloped areas in the use of lime. For example there seems to be no general agreement on the ratios that should apply. Communities surveyed felt that they need to look at demonstration houses to gain confidence in the material.

Like the other industrially processed binder, cement, lime tends to rely on specialized personnel although to some degree owners have been involved in the handling of the material.

There is a distinct aesthetical image that is a result of using lime: white washed buildings with a smoothness to the texture and a high reflectivity of light stand out in whichever context they are in. Cement/Lime mortars tend to offer a high level of satisfaction. In the Koru/Muhoroni area where these were prevalent, all respondents indicated the current quality as 'satisfactory' or better.

Factors influencing access to binders

This section evaluates the various factors which influence the use of binder in the cases surveyed. The binder alternatives covered are cement, lime, cowdung and special earths.

Availability and cost of the binder material

It is necessary that a binder be readily available within the community in such a way as to ensure easy procurement. Indigenous technology-based binders are often widely available among the communities who use them. Procurement of such binders involves simple collection, or 'harvesting' methods almost at zero financial cost. Distances involved also tend to be relatively short such that human or animal transportation is often adequate, thus minimizing on transport costs. Cowdung and special earths constitute some of the key binders in this category.

Table 53: Binder-use constraints cited in case studies

Case study	Factors cited	% of survey
Kinyago village	a. High cost of cement and construction	88
	b. Materials requiring no binder	12
Passenga Settlement	a. High cost of cement	100
	b. High transportation cost	75
	c. Need for constant maintenance for cowdung	25
Majengo-Nyeri	a. High cost of cement	100
	b. Lack of tenureship on land	25
	c. Distance to site and labour costs of special earths	13
Majengo-Machakos	a. Lack of money/high cost of binders	100
	b. House is temporary and does not warrant use of binder	17
Koru/Muhoroni	a. High cost of cement	80
	b. Temporary house, does not warrant use of binder	40

On the other hand, industrially processed binders like cement and lime are manufactured in locations which are far off from points of use. This requires a complex marketing and distribution system which in turn inflates the cost of binders. Transportation, handling and storage expenses become major components of the cost which the house owner or builder pays for the binder. The supply of such binders is prone to occasional disruptions arising from malfunctions in the production or distribution systems and networks.

The availability of a binder over time is also critical in determining a community's access to binders. Besides the essential need for regularity in supply, there is the more critical factor of long-term availability; the issue of sustainability: can the community expect to continue getting normal supplies of the binder at reasonable costs in the next ten, twenty or even fifty years? If the answer to such a question is 'no', then access to the binder will be gradually impaired over time.

Availability and cost of cement

Cement is produced in two large plants: at Bamburi near Mombasa on the coast, and at Athi River, near Nairobi. The Bamburi factory is by far the largest and is the predominant supplier to most areas of the country, apart from the centre. Cement is reasonably easily available in most places and was in fact the most commonly used binder in Kinyago, Nairobi as well as Koru and Muhoroni. The cost price of cement in the various major towns is given in Table 54. To this, one needs to add a transport cost, of usually 10 KSh/bag. Its cost in the survey locations was roughly similar.

In most of the houses where binders had been used, only the external wall surfaces had been plastered; less than half had the internal walls rendered as well. The high cost of cement was a key factor for home owners in deciding to phase the plastering of wall surfaces, and to begin with the most visible and exposed ones. Further evidence of the importance of the cement price as a constraint to housing improvement is given in Table 53. In 5 out of 7 locations, 80% or more of the owners cited the high cost of cement as a key constraint. The inavailability of cement, on the other hand, was not considered a critical factor, except for the fact that it increased its transport cost.

Availability and cost of lime

Unlike cement which is centrally produced in two plants within the country, lime has a more decentralized pattern of production. Two production and distribution patterns are evident:

- locally-based production centres which rely on indigenous technologies of small kilns and woodfuel as the source of energy.
- centralized plants using intermediate technology, characterized by labour intensive methods and medium scale operations.

Lime does not percolate as far down in the distribution network as cement. Its availability is limited to hardware shops in major towns with railway stations or the ones close to such towns. Poor demand in the informal sector of the construction industry could be the main reason for the unavailability of the binder at the smaller centres. Unavailability and lack of awareness of lime as an alternative binder and its performance were the key factors which limits the use of the material. The approximate cost of lime at various major towns in the country as at May 1994 are shown in Table 54.

One tonne of lime is twice the volume of 1 tonne of cement. So in actual use, when batching by volume, 1 tonne of cement is equivalent to 0.5 tonne of lime. Therefore, when mortars or plasters are batched by volume, lime appears to be cheaper than cement, and thus more affordable to most house builders. But, since the strength of lime is less than that of cement, to get a comparable quality of lime mortar to that of cement mortar, the batching should probably be based on weight and not volume. The apparent price advantage of lime over cement becomes negligible when the mixes are based on weight.

Table 54: Cost of lime and cement in major towns of Kenya (May 1994)

Town	Cost of lime KSh $/Tonne	Cost of cement KSh$/Tonne
Nairobi	8 614	8 160
(white lime)[1]	$157	$148
Mombasa	6 140	8 400
(white lime)	$ 112	$153
Nyeri	9 600	8 400
(white lime)	$175	$153
Nakuru	6 260	8 160
(grey Lime)	$114	$143
Kisumu	6 000	9 200
(grey Lime)	$109	$167
Eldoret	6 580	8 160
(grey lime)	$119	$148
Kisii	6 600	9 600
(grey lime)	$120	$175
Kericho	6 000	9 000
(grey lime)	$109	$164

1. Grey lime cheaper by almost 1,000
Source: Homa Lime Company Ltd

Availability and cost of cowdung

Dung is always available among communities which rear cows, and for a long time it was available for free. Inspite of this major advantage, the availability of cowdung as a binder is limited by three major factors, these are:

- the material can not be distributed over long distances: a week or two is the maximum storage period.
- unavailability in urban areas and also for other communities who do not rear cows.
- competition from its other uses as manure or fuel.

Due to the three constraints on availability, cowdung is only used in very few areas, and in communities where it is culturally acceptable. The economic value of cowdung-plastered houses is often viewed as very low.

Availability and cost of special earths
Special earths are mineral-based and indigenous technology-based binders which are available in several parts of the country. They are usually found in small isolated pockets just like clay, but in a wide variety in terms of texture, colour and quality of bonding.

Since they are non-processed binders (usually used in their raw form), special earths can only be used in neighbourhoods close to their source. However, unlike cowdung, special earths can be distributed over longer distances because they are not perishable. Further availability of special earths is threatened by the resultant environmental degradation.

Availability and cost of complementary materials
This constitutes another key factor which influences access to binders in the case studies covered. The key complementary materials for the most commonly used binders are as shown below:

Table 55: Schedule of complementary materials to various binder alternatives

Binder	Complementary materials
Cement	Sand or quarry dust, water (lime optional)
Lime	Sand or quarry dust, water (cement optional)
Cowdung	Ash, water (sand or sawdust optional)
Special earths	Water, sand (optional)

Complementary materials to cement or lime
Sand is the main material required for mortars. Although it is widely available throughout the country, it is only quarried in a few locations and then transported in bulk to the various construction sites. In most cases, the cost of transportation constitutes the largest component of the overall cost of sand. Assuming a ratio of four wheelbarrows (approximately one tenth of a ton) to a bag of cement, this adds an extra cost of approximately KSh 100 to the 50kg bag of cement.

The availability and cost of sand as a complementary material in plasters and screeds for low-income housing is threatened by:
- high competition with the formal sector of the construction industry which can afford to pay higher prices for available sand.
- rapid rate of environmental degradation at the source areas, hence the need for resource and environmental conservation.
- rapid increase in the cost of fuel and servicing of vehicles which leads to frequent escalations in transportation cost.

Quarry dust is often used as a substitute for sand due to its cost-saving advantage. This arises from the fact that the material is procured as waste from close by quarries, and hence involves minimal transportation costs. It can also be procured in smaller quantities from the source using pick-ups or manual transportation. Future availability of quarry dust is closely linked to the future of quarrying especially for building stone in view of the environmental degradation involved and the need for resource and environmental conservation, as well as its alternative uses, e.g. as pozzolanas in PPC.

Cement or lime plasters and screeds require intensive application of water during the mixing and curing processes. The availability and cost of water were noted to be major determinants on the quality of plasters and screeds. The limited availability and high cost of water often lead to inadherence to the normal practice of plaster and screed application in terms of mixing and curing thus hampering the quality of the resultant surfaces.

Complementary materials to cowdung as a binder
Ash and water are the key complementary materials for cowdung. In some cases, owners and builders opt to use sand as a substitute to ash.

Within the localities where cowdung/ash plaster is commonly used, ash has been abundantly available almost free of charge. However, in the recent past (about 5 to 10 years) both cowdung and ash have become relatively scarce due to population growth, decline in fuelwood supply and lower ash production as a result of changing cooking habits and cultural changes. Thus there is an emerging trend of commercialization of these materials. To overcome the scarcity and pending commercialization of ash, some owners and builders in Gititu opt to use sand as a substitute. However, this alternative has not been widely applied and therefore its full potential versus that of ash is yet to be established.

Since sand cannot be locally procured like ash, it inevitably involves procurement and transportation costs which would not be significantly different from those of sand used for cement-sand plasters and mortars.

Water is widely available, normally from nearby rivers.

Availability and cost of application technologies
The indigenous technology-based binders rely heavily on widely shared application skills and non-specific tools. Such skills come very close to what is termed as traditional knowledge or skills in that almost any mature human being in the community can actually do the work, although there

might be a division of labour based on gender, age, or social status in the community.

The application technology for cement requires highly specialized skills and tools. To acquire the skills, most of the artisans interviewed had gone through the apprenticeship process under experienced artisans. It takes a period of 2 to 5 years of apprenticeship before the apprentice starts working independently. The specialized tools for application of cement plasters and screeds include the plumb, spirit level, float, trowel and 'gurumala', among others. The cost of the tools require a capital outlay. Artisans who cannot afford their own tools have to borrow them from their peers in the trade.

The hiring of artisans (fundis) for cement application ranges between KSh 80/day in some rural areas to KSh 200/day in urban centres like Nairobi. Labourers wage range between KSh 50/day to KSh 100/day. The lowest combined wage for 'fundi' and labourer comes to about KSh 130/day, while the highest could get upto KSh 300/day ($5.5). In all the case studies, community leaders and owners of houses expressed the opinion that the cost and unavailability of fundis can be a severe constraint to the use of cement.

Although the tools required for the use of lime are very similar to those of cement, findings in the case studies show that the culture of lime-use is different from that of cement. The culture of lime use requires a different treatment of the material in terms of 'maturing', mix ratios, setting time, curing requirements and finish of surfaces. Lime is also viewed as a highly corrosive material to the hands, thus requiring to be handled with care. It can be observed that skills of lime-use are not readily available. However, most of the fundis who use cement can be trained in the use of lime for faster dissemination of the skills.

Cultural acceptability of binders in the case studies

Access to a binder also depends on the cultural acceptability of the material. This depends on two major factors which are:
- general awareness of the availability and performance potential of binder alternatives.
- socio-cultural values associated with the use of a particular binder.

A community's access to a binder is greatly enhanced where there is widespread and accurate information concerning availability of the binder and its performance potential, and when at the same time the use of such a binder is associated with positive socio-cultural values. A binder which meets these two criteria inevitably achieves high cultural acceptability and it becomes popular. If a binder scores negatively in any one or both of the two criteria, then its cultural acceptability level is very low, and it is unlikely to be widely used.

Conclusions and recommendations

Access to binders and the art of binder use is limited because of the variety of binders available on the market. The available binders present different qualities and characteristics which mitigate against their wider use. The high performance binders-based on high technology - have had greater social acceptance but have been lacking in affordability, availability and flexibility of use. Economy is therefore a central factor in the use of high-tech materials like cement. While traditional materials like cowdung or special earth allow for the full application within the building, barring certain technical factors, there is a tendency to limit the use of cement either to walls or just to the external side of the walls in order to economize on cement expenditure.

The indigenous technology binders face their greatest obstacle from continued availability (which is uncertain) and social acceptance. There is competition for the use of cowdung and commercialization has crept in, underlining this developing scarcity. With increasing environmental awareness, the mining of special earths is unlikely to continue unabated.

The economic factor, alongside cultural acceptability, is going to be a determinant in the continued access to binders and the variety possible. The apparent search and consequent acceptance of a hybrid between high technology based binders and indigenous binders offers one indication of the future. Intermediate technology binders begin to suggest themselves as this option. They will need to combine the best characteristics from both the high technology and indigenous technology based binders. Underlying their acceptability is the issue of quality. The quality of the resultant work must be sufficient to address the issues of concern to the builders and owners. These are tested mainly in the longer term when weathering, and general deterioration sets in.

It is also clear that the diffusion of a binder-use culture will have to rely on acceptance at a communal level. A degree of confidence among the community members and artisans at this level is a necessary prerequisite.

It will be necessary to increase the variety of binders on the market, the largest thrust for further development being the intermediate technology based binders. Lime has a central role to play in this respect. Another avenue of increasing binder variety is to produce blends. This could be done by starting with the binders already in use e.g. earths stabilized with lime or cement. There is also great potential for binders blended with pozzolanas. These new alternatives should be introduced to the users together with their application technology. The builder (fundi) needs to be trained so as to have the necessary technical skills of handling the various types of binders.

Taken together, these factors point to the need for some form of intervention in the binder availability. It points to a need for an evaluation of technological levels with a bias toward the intermediate technologies. The acceptance of hybrid material among the communities encountered in this study may provide a strong and persuasive base to design an intervention strategy. This approach may require that a system of incentives be instituted to attract the private sector to expand their involvement. Such incentives built around market expansion strategies, could be part of a wider program geared toward the introduction of relevant skills and awareness enhancement on matters to do with binder use.

An approach to positive intervention must be geared toward increasing the alternatives available to builders rather than trying to substitute the available choices. This will mean taking cognisance of the prevailing building culture and working within it. There is a need to explore and initiate dissemination programmes that could expand the range of choice of binders within low income housing communities. Within this training system, a wider base for communicating with the wider community needs to be created. Demonstration units are a proper approach as they serve to instill confidence in a material by providing 'eye-evidence'. It is also critical to work with recognized institutions like youth or women's groups. Such institutions provide a base for shared realizations and apprehensions and are therefore important contact points in society. It is moreover necessary that those who have this new information be easily accessible and approachable. A system of 'extension workers' on hand to offer suggestions and answer questions in the knowledge of local conditions would be ideal. This suggestion carries the seeds of a wider thought. What may be required is a new type of professional architect specifically trained to work with communities whose demands are different from those whom the conventional architects are trained to handle.

Underlying the efforts to introduce the new materials must be an ample supply of patience and a constant recognition of the need to work closely with the community and especially the artisans. Self-sustaining confidence in new materials is likely to take time building due to the need to counter prevailing attitudes. Research into the enhancement of the performance potential of the indigenous technology based binders through blending with cement or lime would be a worthy starting point. Such research would also be geared towards enhancing the cultural acceptability of such binders and their future availability.

References:

Cowie, A. P. (ed.) *Oxford Advanced Learners' Dictionary*, (4th ed.). Oxford University Press, Oxford, UK, 1992.

Denyer, S. *African Traditional Architecture*, Heinemann Educational Books Ltd., London, 1978.

Fathy, H. *Architecture for the Poor*, University of Chicago Press, Chicago, 1973.

Jayawardene, S.:"Reflections on Design in the Context of Development", in *Mimar 27*, March 1988, pp 70-75, Concept Media Pte Ltd., Singapore.

Khan, H. and B. Taylor, "The Development Workshop", in *Mimar 1*, 1981, p.37, Concept Media Pte Ltd., Singapore.

Low, M. S. and E. Chamber (eds.), *Housing Culture and Design. A Comparative Perspective*, University of Pennsylvania Press, Philadelphia, 1989.

Norburg-Schulz, C. "Genius Loci: Towards a Phenomenology of Architecture", Academy Editions, London, 1980.

Oliver, P. (ed.): *Shelter in Africa*, Barrie & Jenkins Ltd., London, 1971.

Ortega, A.: "Basic Technology: Lime and its Production", in *Mimar 17*, July - September 1985, pp.77-81, Concept Media Pte Ltd., Singapore.

Ortega, A.: "Basic Technology: Gypsum, its Production and Uses", in *Mimar 18*, October - December 1985, pp.58-61, Concept Media Pte Ltd., Singapore.

Rapoport, A.: *House Form and Culture*, Prentice Hall Inc., New Jersey, 1969.

Schumacher, E. F.: *Small is Beautiful: Economics as if people mattered*, Harper & Row, New York, 1973.

Smith, R. G.: "Alternatives to OPC", *Overseas Building Note (OBN)* 198, March 1993. Building Research Establishment (BRE), Watford, U. K.

Tuts, R.: "Potentialities and Constraints for using Pozzolanas as Alternative Binder in Kenya", paper presented at the First International Seminar on Lime and other Alternative Cements, 9-11 December 1991. Housing Research and Development Unit (HRDU), University of Nairobi, Kenya.

Tuts, R.: "Pre-Feasibility Study on the Use of Rice Husk Ash as Cementitious Binder in Kenya", in *A Study on the Present Situation and Proposal for a Development Strategy of the Cement Industry in Kenya*, UNIDO, 1992.

Wenzel. M.: *House Decoration in Nubia*, Gerald Duckworth & Co. Ltd., London, 1972.

Uganda

B.M. Kiggundu, B. Nawangwe and D. Tindiwensi

In defining a need for binders or quality improvements and durability in housing units, it is important that basic surveys are made. Thus, the present effort concentrated on the use of binders in low-cost housing construction. In addition houses built without binders were studied for comparison.

The survey efforts focused on both urban and rural settings and the result uncovered the following:

- affordability of binders by low income communities.
- access problems faced by would be users.
- types of binders selected and associated purpose.
- what technologies are used in binder application.

In addition, the survey findings indicated some future market potential for alternative binders, technologies of use, and guidelines for a dissemination strategy.

Case studies

Kanara Dura river

Kanara trading centre is situated about 10 km from Kamwenge town towards Kasese. Kamwenge is about 350 km from Kampala along the Kampala Kasese railway line. The village is a typical example of remote parts of Uganda with no access save for the single feeder road which is being opened by the German Technical Co-operation (GTZ). About 6 km from the village, after the Dura river, there is a lime factory run by the production unit of the National Resistance Army (NRA). Cement is produced about 80 km away at Hima, by the Ugandan Government.

From the interviews carried out, it is evident that:

- The costs of building materials such as cement are too high for the rural peasant to use. This has prompted them to resort to temporary structures.
- There is a perennial problem of termites which destroy houses in this village and its surroundings.
- Until recently, transportation constraints have denied the residents of this village and its surroundings the use of better binders produced elsewhere without standing higher costs.
- There is an acute lack of water in the whole of Kibaale district.
- Seismicity in this region is another constraint for better housing. The recent earthquake destroyed masonry structures whereas traditional mud and pole houses survived. This has left people to wonder whether traditional mud and pole houses are the best suited for this environment.
- There is a general lack of practically oriented youth who could develop the technology of blending binders with other materials to improve low-cost housing.

Where binders had been used in this village, a quarter had used lime and the remainder cement; no cases of using pozzolanas or special earths were noticed. The reasons for this use of binders put forward by the residents are:

- Cement and lime are the only available binders in this part of the country.
- The available builders only know how to deal with these binders.
- Other binders like pozzolana and special earths are not known in this village.

Lime is a new product in this area. It is only in 1987 that the National Resistance Army (NRA) production unit revitalized lime production at Dura. Before then the only available binder was cement. Again due to the same reason, the technology of using lime with various base products is not fully developed.

In this village, a wide range of base materials are used in house construction. The table below gives the summary of base materials used in houses. Where binders were used 68% of the surveyed houses were built out of burnt brick masonry, 28% out of mud and pole, only 4% or one house out of concrete blocks and no house was built out of mud blocks.

This shows that in this village the technology of making and building with burnt bricks is well developed. However this is expensive for peasants. This is the reason why almost all the houses where this material was used were found in the trading centres of Kanara and Kamwenge. However houses where mud and pole provided the base materials show that:

- traditional materials (mud and pole) are compatible with modern ones (cement and lime).
- this combination of materials is cheaper than using burnt bricks; mud and pole houses can be found deep in the villages.
- the technology of combining these materials is well developed in this village.

It is interesting to notice that of the surveyed houses, none was built out of sun dried mud blocks. The obvious reason is availability of forests in this part of the country which provide wood used in baking the bricks. It is also important to note that there is an acute problem of water which puts the cost of a mud block house at a relatively higher level.

Table 56 compares wall base materials with binder use. These results provide clear evidence that the type of base material influences the choice of the binder. The exact relationship between the two cannot be determined here but the evidence is clear enough to make this judgement. It is also very important to note that much as the compatibility of lime and mud influences their being used together, the effort of keeping costs low should not be under estimated since lime is much cheaper than cement in this village, at USh 2,200 (US$2.2) per 25 kg bag, versus USh 11,000 ($11) for a 50 kg bag of the latter.

Table 56: Binder use in Kanara village, Uganda

Base materials	Cement (%)	Lime (%)
Brick masonry	17 (71)	3 (37)
Mud blocks	0 (0)	0 (0)
Concrete blocks	1 (4)	0 (0)
Mud and pole	6 (25)	5 (63)
Total	**24 (100)**	**8 (100)**

The above is an indication that cement tends to produce better finishes than lime in the village. This can partly be explained by the fact that lime is a recent product of the area and therefore the technology for its use especially in buildings is young. Secondly, all the technical institutions in Uganda which train builders base themselves on cement. Little is known and taught about alternative binders. This leads to artisans who are ready to do a good job with cement but unable to do so with other binders. This disparity also partly explains why cement is widely used despite its being more expensive than lime.

The category of houses where conventional binders were not used is made up of traditional houses. Kanara village, which is found in Kibaale district is largely occupied by immigrants from Kabale district, in South Western Uganda. This is a Bantu tribe whose traditional architecture involves mud and wattle houses built on a square plan. The materials used include: mud, eucalyptus poles, elephant grass reeds, spear grass, silt, sand and cowdung. Cowdung is used as a natural binder. It is mixed with sand/silt and water and the resulting suspension is smeared onto walls as plaster. Most houses had good structural designs with appropriate slopes but the finishes were poor. Most respondents in this category explained that:

- the cost of building a permanent structure was prohibitive, thus forcing people to resort to traditional homes.
- the problem of termites in the area which destroy virtually all the wooden components of the structures, had a negative impact on better housing.

Therefore these structures are built to last a maximum period of ten years, after which they are replaced.

Tororo municipality

The town is 120 km east from Kampala within Bukedi province. It is the third largest town in the Eastern region and accessible by both road and rail transport. Due to peace and security prevailing today, the inhabitants are trying to improve the structures thus increasing the sophistication and size of the town.

Tororo is well known for its limestone. Production units have been set up to exploit these resources. Of the resulting products, Tororo is more known for its cement than for its lime.

The following factors influence shelter construction in the town:

- the costs of building materials are high and only high income earners such as businessmen can afford moderate housing facilities. Normally, rental houses are built, hence most of the residents are not necessarily home owners.
- transport costs aggravate the picture since for some materials like bricks, the transport costs may be double or triple their purchase price.
- most of the semi-permanent and temporary structures are occupied by either peasants or low income earners.
- some of the semi-permanent structures are the result of an enforced requirement from the Municipal council. Some of the residents in the above mentioned structures intend to put up permanent ones as soon as the land is surveyed.
- it was disclosed by the local authorities that all the people in the area have a bias for cement as it is the only strong binding material. So there is need to educate them about other binders.

Despite the aforementioned, the people of this town are continuing to build but at a slow rate because of the exorbitant costs involved.

Of all the houses surveyed where binders had been used, only one had used lime, but all the others cement; and even the use of lime was mixed with cement. The reasons for this uniform trend were given by respondents, namely:

- cement has a high durability status and is long lasting.
- cement is available at the hardware stores and relatively cheap since the manufacturing factory, Uganda Cement Industry is nearby.
- because of the proximity of the factory, the material is accessible

At this point it is important to understand why only one house out of the twenty five visited used lime. The reason given by most respondents was that cement can sufficiently act as a binding material on its own, while lime can not. The other reason basically reflects what has been mentioned earlier in reference to the durable nature of cement. Most home owners feared that lime was a weak binding material and not suitable for permanent structures. However, the one home owner that used lime wanted to save costs. Lime is sold in Tororo at USh 2,000 ($2) per 25 kg bag, whereas a 50 kg bag of cement costs USh 7,500 ($7.5).

Of the 25 houses visited in Tororo, 16 used brick masonry and 9 concrete blocks. This result reflects the urban nature of the area where people use modern building materials as opposed the traditional ones. The reasons given for an increase in the number of brick masonry houses in recent years are:
- before 1980 there was little knowledge about the technology of making bricks.
- concrete blocks are much more expensive than brick masonry.
- following this recent innovation, bricks are easily manufactured and this can be done locally.

On the other hand, mud blocks are not used in the municipality because they do not have reasonable strength and also most of the area is rocky hence soil is a scarce material.

The quality of the finishes of the surveyed houses is summarized in Table 57. Here again the presence of cement as a binding material plays a crucial role.

Table 57: Quality of finishes in Tororo, Uganda

Feature	Very high	High	Medium	Poor
Floor screeds	12	6	5	0
Wall plasters	11	8	2	2

The factors influencing quality are:
- *workmanship:* bad workmanship leads to the cracking of floor screeds and wall plasters due to poor mixes of the mortars used.
- *age:* the deteriorating nature of floor screeds due to their being subjected to wear and tear will after some time require repairs.
- *weather:* this element interferes mostly with exposed wall plasters.
- *poor handling:* occupants tend to mishandle the structure leading to dilapidation.

Ten houses were surveyed, where conventional binders were not used. Most of these represent traditional architectural houses and some are largely low cost structures but with modern designs. The materials used are those which the local people could easily get access to in their environment. These include: mud, eucalyptus or Musizi poles, elephant grass reeds, spear grass, sand, cowdung, and molasses and ash for plastering. Cowdung is used as a natural binder. It is mixed with molasses, sand, and ash and the resulting mixture is manually smeared on the walls as plaster. This type of plaster however requires constant maintenance where fresh coats are added after every 3 years. Most of these buildings would last for a good period of time but due to neglect by the occupants, the houses were in a pathetic condition. Despite this shortfall, the owners supported such structures for the following reasons:
- the low costs of building them
- temporary structures suited very well the situation of temporary land ownership. The local authorities refused people setting up permanent structures on land which is not duly surveyed and correct ownership determined.

These low cost houses may have served as ideal shelters for the low income earners but Tororo being an urban area, the authorities require that permanent structures be built. This they claim, is to reduce the number of unnecessary renovations and maintenance. However a blending of the two technologies to provide an intermediary may perhaps be the most desired solution.

Namayumba

Namayumba is found in Mpigi district and located along Kampala-Hoima road approximately 29 miles from Kampala, the capital city. It is linked to Kampala by a tarmac road. Thus it can be argued that it has good access. The community is predominantly rural, and practices subsistence farming. But it can optimistically be pointed out that a tradition of urbanisation is slowly creeping into the community.

A survey about the use of binders in low-income housing was carried out in this area. On discussion with the local leaders and the residents concerning housing in the areas, the high costs of cement and transportation were the major constraints cited. Another problem was the effect of the civil war which raged in the area for almost a decade. Despite these setbacks the local leaders are optimistic that with increasing income for the rural population, reasonable housing will increase in the area. This is because:
- there is a general tendency among the residents to move to better housing with increased income
- the area has skilled artisans capable of backing up the construction process if need were to arise.

Namayumba happens to be in an area that was subjected to the effects of war for quite a long time. To this end, a large number of the sample houses chosen were falling within the age range of 2-6 years existence.

Amongst the houses where binders had been used, there were two cases of the use of special earths; of the remainder about one third had used

lime, and the remaining two thirds cement. The reasons given for the wide use of cement are that:
- it has a fast strength development which makes construction work to be quicker and easier. This explains the wide use of cement for masonry work.
- the binder allows the development of a hard and strong structural components which do not appreciatively deteriorate with time. This property is widely taken advantage of in constructing the floor, a structural component that is subjected to varying loads.
- the product in which cement is used as a binder can withstand extreme conditions such as exposure to rain without serious deterioration.

The relatively lesser use of lime as a binder is due to the fact that the sand used contains substantial quantities of clay which has lead to poor performance. Because of peoples' failure to identify the cause of the poor performance, it is now an accepted fact is some quarters that lime is a poor binder. The other factors that contribute to the low use of lime are the longer setting time as compared to that of cement which results in the slow progress of construction work and the low strength development as compared to cement. In order to solve this problem, the builders mix lime with cement. This, according to them, produces a stronger mortar and has an early strength development. The use of a lime cement mixture also has the desired effect of reducing the costs that would be incurred if cement alone were to be used.

The non usage of pozzolana as a binder is due to the fact that it is not yet available on the market. However, after an explanation about pozzolana, and the effect it can have in the construction industry, there was enthusiasm shown by respondents to acquire it in the quickest time possible.

A binder considered under the special earths is cowdung. This when mixed with silt, and water is added, has a performance that closely parallels that of cement. The greatest drawback of special earth, more in some localities than in others, is their inability to withstand the effects of rain.

The ten sample houses in which no binder was used have for their constructional material, mud blocks or burnt clay bricks held in place by earth or earth walls reinforced with reeds, and compacted murram mixed with mud for the floor. The reason advanced for not using binders by the owners of these houses is the high cost of cement which is normally beyond their economic reach. But because of the fact that such houses pose problems arising from dust, offer a good habitat for small insects such as flees and are liable to attack by termites, have low resistance to rain etc., the owners argue that any slight opportunity for building using binders that presents itself, is eagerly seized upon.

Burnt clay bricks are a very common wall material in this area, becaue of the availability of materials and labour, low costs, and simplicity of brick manufacture and construction. The local builders further argue that the high compressive strength of burnt clay bricks, their good resistance to rain, moisture, and insect attack and the ease of transporting them from place to place without major losses make them cherished as ideal and economic building materials in the area. The wide usage of brick masonry can also be attributed to the fact that the bricks to be used for construction are prepared and burnt on site. This has an effect of reducing construction costs.

The low compressive strength, susceptibility to erosion by rain, general loss of stability on absorption of water, low resistance to abrasion and their vulnerability to damage by termites explain why mud blocks are not widely used despite their being cheaper to produce as compared to burnt clay bricks. The use of concrete blocks is very minimal. It is argued that their use would result in extra construction costs incurred in their production and use.

On the other hand houses built out of mud and pole can be amazingly durable in some localities in addition to being cheap. These houses represent man's desire to have comfort that binder usage can generate but at the same time saving in construction costs. It can be pointed out that such a house is an upgraded form of the traditional mud house. Their major drawback is the difficulty of keeping the walls plumb and straight which results in the finished walls having depressions and therefore being aesthetically unpleasant. This base material is also liable to attack by termites. The different rates of expansion of mud blocks or the mud in mud and pole houses and cement plaster which causes the latter to peel off also contributes to the limited use of mud blocks and the mud and pole base materials.

Within this community, there is a desire to have modest houses built out of brick masonry using binders because of the comfort generated. A clean house environment, no problem of insects like fleas whose habitat is dust, less liability to attack by termites, and durability. This argument explains the wide use of cement and lime with brick masonry.

An inspection of finishes reveals that a large number of sample houses in the area had wall plasters and floor screeds that are of high quality. This result can on the whole be classified as good and can mainly be pointed to the fact that the local builders have good experience. However the medium and poor occurrences can be explained thus:
- poor mix ratios mainly due to the insistence of the clients which results in weak structures.
- poor designs which do not allow easy movement of people and objects in the house. This results in

scratches on the finishes thus increasing the rate of deterioration.
- it is clear from the findings that lime is not used for making the floor.

The reasons given for this are:
- it has a low strength and since the floor is liable to be subjected to varying loads, the use of lime in floor screed construction is discouraged.
- lime has a longer setting time than cement and therefore more time would be needed to construct a floor area than if cement were used.

Cement is the most widely used binder in making the floor screed. The performance of cement as a binder in this area is appreciated by the owners of the sample houses. An examination of wall plaster quality in relation to the binder used among the sample houses revealed the results shown in Table 58.

Table 58: Quality of finishes in Namayumba, Uganda

Binders	Very high	High	Medium	Poor
Cement	3	15	3	1
Lime	1	6	2	1

Almost all the owners of the sample houses obtained the binders from Kampala. It is generally agreed that the binders are easily accessible upon deciding to use them. The average cost at which the two most commonly used binders in the area were obtained on the market are cement USh 11 000/bag, lime USh 4500/bag. The cost of cement is high according to the residents. This is reinforced by the fact that an overwhelming majority of people interviewed cited the high cost of cement as one of the major constraints met during the construction of the house. Because of the alleged high cost of cement the binder most commonly used, there are three phenomena that were noted in this area:
- People blend traditional and modern building materials when they are constructing their houses. A case in point is the use of Misambya tree trunks that find respect among traditional African builders for its strength, durability, and resistance to insect attack. In a bid to reduce the construction costs, the Misambya poles are used to make the roof trusses and purlins to support the iron sheets.
- Structural components, namely lintels and ring beams which require stronger mixes and consequently use of more cement and hence higher construction costs, are left out during construction. This usually leads to cracks especially at door and window joints but the fact that the load arising from the iron sheets is light makes these houses to stand for quite some time.
- People live in houses that are not plastered for quite a long time. If plastering is to be done at all, the internal parts are given first priority.

There was no complaint about the quoted price of lime. Despite the alleged high cost of cement, the mood of most local people here is to have a modest house built using binders.

Ezuku village - Arua

Ezuku is an area where no lime or other binders are produced, with bad road access. It is located near the Uganda-Zaire boarder. The ethnic group living in this area are the Lugbara.

Where binders were used, this was predominantly cement, in two-thirds of all cases; there were a few cases using a mixtures of cowdung and soil, and some more using lime-cement mixtures, which was a recent innovation limited to houses built in the last five years. Some reasons advanced for this distribution were:
- cement is the only known binder in this part of the country in which people have confidence.
- building lime is not popular because little is known about it and people have less confidence in it.
- other binders like pozzolana are not known.
- cowdung-sand-ash mixture is considered as having inferior and poor properties and hence a poor man's option.

The quality of finishes was classified as very high, high, medium, and poor. From the findings no floor screed could be classified as very high in quality. The majority were of poor and medium quality categories. Wall plasters exhibited all the category criteria and it was found out that 32% were poor, 37% were medium, 21% were high and a smaller 5% were very high in quality. Poor plaster finishes mainly showed on walls built out of mud and pole, mud blocks, and those exposed to rains. The majority of poor floor finishes were exhibited by floor screeds laid directly on compacted soil where no hard core, concrete or proper base material was in place. Poor mix ratios and techniques also contributed. Very high and high quality floor screeds and plasters were mainly found in houses built by local artisans who once worked on construction sites with Indians or colonial masters.

The ranges of costs between 1990 to 1994 for cement were USh 11,000-15,000 ($11-15) per 50kg bag, and for lime USh 5,000-6,000 ($5-6) per 25kg bag. Fluctuations in prices are not very significant.

Ten cases of traditional houses in which no commercial binders were used were considered. These had either walls and foundations out of adobe blocks and mud masonry or of mud and wattle, and a thatched roof. Wall finishes encountered included sand-ash or sand-ash-cowdung plasters which set like cement-sand plaster, with a final coating of black valley mud and decorations of white, brown, red and black soil. The floors of thoroughly compacted earth are regularly smeared with cowdung which may be mixed with soil or local brew residues.

Both circular and square houses were encountered. It was noticed that the circular houses were built as kitchens and the latter were main houses. Many of the respondents expressed the wish to build with commercial binders because:
- termites were a major problem eating poles, grass, and building mounds inside houses.
- the houses required too much attention with respect to maintenance.
- the houses did not last and in case of fire one saves nothing.
- it was difficult to build beyond a certain size. But the main problems expressed were that:
- commercial binders were not affordable
- binders are scarce in this part of the country.
- transport to site may be difficult and costly to undertake.
- a variety of binders is not available for one to choose cheaper options.
- many people have never thought of building using commercial binders because a carefully designed combination of local binders could produce a good looking and comfortable house.

From the survey, the will and wish to build using commercial binders are alive but financial, technical, accessibility, availability, and variety constraints limit their popular use and application amongst the low-income rural community in the area.

Kabaawo village - Mutundwe parish, Kampala
Cement is the most popular binder used in this locality, with 61%. Lime is used in the remainder of the cases, but pozzolana has never been used. Probably it was unknown to this community and non-existent in the market at the time of survey.

Table 59: Binder usage in relation to base materials in Kabaawo, Kampala

Base materials	Cement	Lime
Brick masonry	15	12
Mud blocks	2	1
Concrete blocks	7	2
Mud and pole	1	0
Total	25	15

As shown in Table 60, cement is adopted for use in all structural components in Kabaawo. Lime is mainly preferred for internal wall surfaces and little used for joints, floor and external wall surfaces. Brick masonry is the most popular base material used followed by concrete blocks. Mud blocks and mud and pole are least used. In all probability, the trend is to move away from traditional construction materials. The price of cement in Kabaawo is USh 10,000 ($10) per 50 kg bag, whereas it is USh 3,000 ($3) per 25 kg bag for lime.

From the analysis one gets the impression that the technology of application of floor screeds and wall plasters in this locality is well developed; that is why very high quality of finishes were observed.

Table 60: Quality of wall plaster in relation to binder used in Kabaawo, Kampala

Binders	Very high	High	Medium	Poor
Cement	8	10	4	1
Lime	6	6	3	0

A further split in the quality of finishes with respect to the various binders gives a similar result to the reading for cement and lime as binders. The people working on these houses are skilled and knowledgeable about the application techniques.

Analysis and conclusion
From the findings, cement is the most widely used binder in all the areas where the survey was based. The results further show that there are only two conventional binders used in the building industry in Uganda: cement and lime. Special binders were found only in Ezuku village. Other binders like pozzolana are not known in most of the areas visited. In fact, no respondent in any area indicated having used pozzolanic cements. However many people expressed the desire to build with binders in a bid to improve their housing, but as seen the biggest deterrent was cost.

The following bar charts (Figures 1-3) compare:
1. the usage of binders in the five areas of survey in terms of percentages.
2. the usage of base materials in the different areas also in terms of percentages.
3. binder costs.

From the rows of averages and their co-efficient of variation one can confidently predict that for all the houses in the surveyed areas in Uganda where binders were used, about 72% used cement while about 66% were built out of brick masonry. However, the wide variation of the other elements do not permit any predictions. This is due to the large number of zeros obtained from the different areas.

The binder cost variation amongst the areas of survey is given the in Table 61.

Despite the high cost of cement in Uganda, people continue to use it regardless of the cheaper lime alternative. This is a clear manifestation that people are trying to improve on their housing conditions through use of the best materials available. There is potential for alternative binders, provided they are made available along with an awareness and education compaign.

However, the existence of unimproved traditional structures even in urban areas is an indication that this attempt is still futile because the average Ugandan is still a poor person who can not afford a reasonable house at the present costs.

Table 61: Statistics of binder cost in Uganda
($1= USh 1,000)

Statistic	Cement (50kg)	Lime (25kg)
Average (USh)	10,500	3,490
Range	5,500	3,750
Standard deviation	800	1,432
Co-efficient(%)	7.6	41

Finally, the (urban/rural) Ugandan poor will remain in unimproved shelter or ill housed if the costs of building materials especially binders do not decrease. An attempt to reduce the cost of binders without equipping the populace with the appropriate technology for their use will not solve the problem either. Therefore as new binders like pozzolanic materials are developed, possibilities of training artisans to use them should be promoted. This can be made simpler by developing binders which blend easily with local materials. In this perspective therefore, local binders like cowdung, ash, and local brew residues are potential starting points.

Note:
A = Kanara trading centre near Kamwenge
B = Tororo municipality
C = Namayumba in Mpigi district
D = Ezuku village in Arua
E = Kabaawo village in Kampala

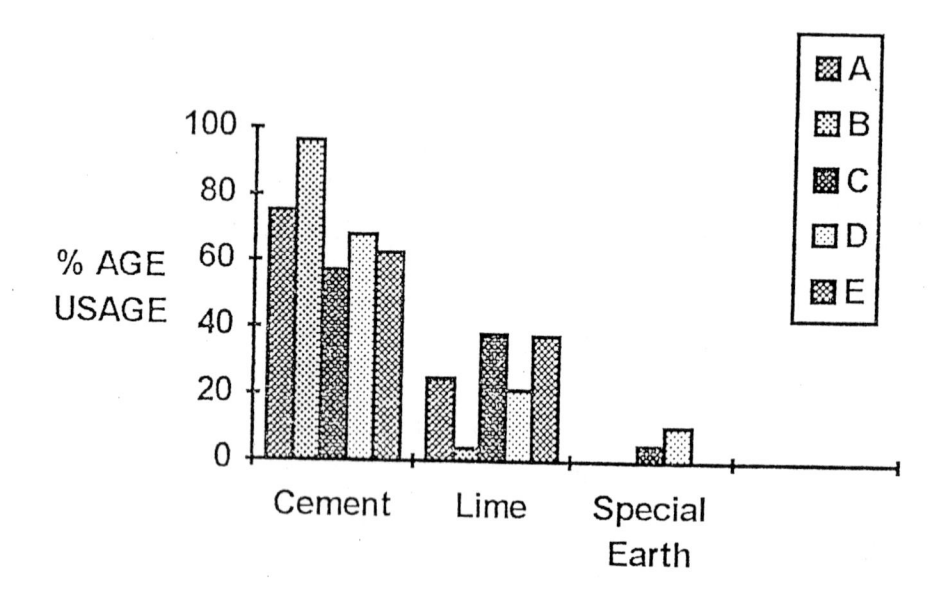

Fig.12: Use of binders in the five survey areas

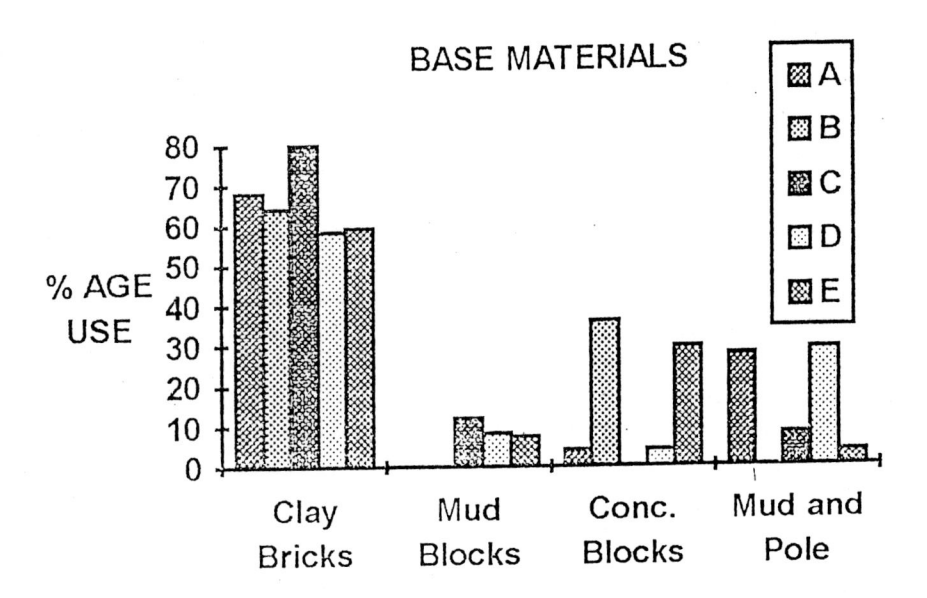

Fig. 13: Use of base materials in the five survey areas

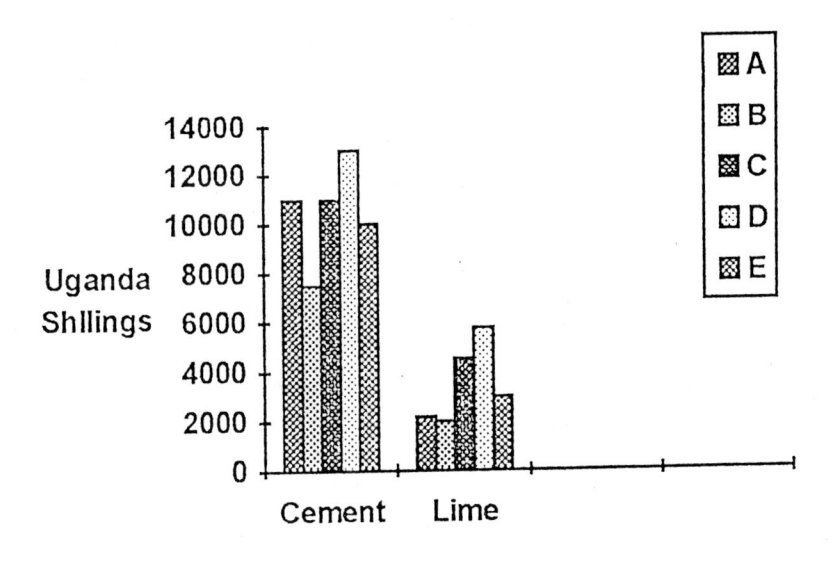

Fig. 14: Binder costs in the five survey locations

Uganda
SECTION 4

Tanzania

Huba Nguluma, Hussein Rajab and Livi'H. Mosha

Promotion of alternative binders is important if low-income earners are ever to afford quality housing. The majority of Tanzanians are low-income earners. The country's population is 25.8 million and its per capita income at only TSh 32,770 (US $60)per year. The purchasing power of the community is very low, thus 23% of the people interviewed did not see the need for improved housing.

Industrially produced building materials are very expensive, and beyond the reach of most of Tanzanians, hence the need for research and promotion of alternative building materials and technologies which can be developed from locally available resources. Tanzania has abundant limestone reserves in all regions except Ruvuma. Thus there is definite potential for lime as a binder for the construction industry in the future.

The survey was carried out in six different locations. The six areas effectively represent the country. Residents and artisans (fundis) of the different houses with or without binders were interviewed. The survey focused on getting information on problems encountered in access to binders, the type of binders they use and the technology applied in using the binders. The study has attempted to expound the need to promote the use of alternative binders. It has explored the advantages of the use of binders and vice-versa. It has for instance been noted that households in which binders have not been used revealed that the bottlenecks which inhibit their use include high cost of the materials especially Portland cement, lack of appropriate technology and a low level of awareness on the use of alternative binders. The majority of people interviewed explained that they are willing to adopt new building materials and technologies and other appropriate materials.

Why does Tanzania need alternative binders? The national shelter policy is in line with agenda 21 chapter 7 adopted by the plenary session of the Rio de Janeiro conference. The right of adequate shelter is emphasized for every human being. To achieve this goal, adequate and quality binders are essential.

Cement is the most common binder used in Tanzania but there are other alternatives like lime, pozzolana, ash and cowdung. However cement production in Tanzania has not matched demand. This means there is a need to:

- rationalize the use of cement
- increase its production to lower its cost
- supplement it by using other indigenous cementitious materials such as lime, pozzolana, gypsum, etc.

Lime is the second most commonly known and used binder in Tanzania. It is important to note that large-scale lime production in Tanzania in 1993 was only 1,139 tonnes according to the Industrial Communities Quarterly Report of June 1994 (ref.3) compared to 748,850 tonnes of cement in the same period. However, apart from the low capacity utilization of large scale lime plants in the country, it is only a very small portion of this which goes to the building industry. Lime is also used in:

- road construction, i.e. as a soil stabilizer
- agriculture- fertility is greatly improved if the pH of certain soils is increased by liming.
- chemicals, food, metallurgical industries
- tanneries
- environmental protection, e.g. softening of hard water

Most lime used in building is produced on a much smaller scale, for instance by heap kilns. In an enquiry made in Tanzania among contractors concerning consumption of lime in buildings, it appeared, that plastering and masonry consume most of the lime. It was also evident that lime technology is not very well known to contractors, let alone the local builders (fundis), with the exception of the islands.

Case studies

Morogoro

In Morogoro, Kihonda village was chosen to represent the villages in which no lime or other alternative binders are produced but have good access roads, therefore stimulating the supply of building materials. Morogoro is located in the centre of Tanzania.

Most houses in this area are constructed from sand cement blocks, mud blocks and very few from burnt clay bricks. The houses of sand cement blocks are almost new because it is a newly developed area. The binder used in this neighbourhood is mainly cement (99%) and only 1% lime. Cement is preferred by most people because they believe that it has more strength than lime. Also the technology of using lime has not yet reached this place. The main problem associated with binders in this neighbourhood is its high cost of purchase (one bag of cement of 50 kg costs TSh 3,000 = $6). This includes the cost of transport (hiring a vehicle), because the cement is purchased in Morogoro town centre.

For the houses where binders have not been used,

they use mud blocks made from the red soil within the neighbourhood. They are generally in a pathetic condition, with leaking roofs and nearly collapsing walls.

Kigamboni

This is peri-urban neighbourhood on the Southern fringe of Dar es Salaam. The easiest way of reaching this place from the town centre is by ferry across the Indian ocean. Most houses in this neighbourhood are constructed from sand cement blocks or mud and pole. Binders used in this area are cement (70%) and lime (30%). Most houses have a cement sand plaster. Lime is predominantly used in painting the walls. Sometimes it is even mixed with pigments to obtain colour paint.

Lime is obtained within the neighbourhood. There are small-scale producers along the sea shore who sell it at retail prices. A 5 kg bag is sold at TSh 150 while a 10 kg bag is sold at TSh 350. Cement is obtained from nearby stores at a cost of TSh 2,800-3,000 per 50 kg bag. Push-carts are the main means of transport.

Houses which do not use binders are mostly built with mud and pole. Soil is obtained within the neighbourhood. Ant hill soil is the best soil for constructing this type of houses. These days it is difficult to obtain poles for construction nearby; people have to walk a distance of about 20km to procure poles for construction.

Oldonyosambu

This village/trading centre is 40 km from Arusha town with a good tarmac road leading there from the town centre. The roads within the village are very poor. The village is very rich in pozzolanic soil. About 90% of village houses are round in plan, made of mud and poles, plastered with a mixture of pozzolanic soil and cow dung. These are the predominant binders in Oldonyosambu village; cement and some lime are used in the trading centre of the area. A 50 kg bag of cement was selling at TSh 3,500 ($7) while that of lime 25 kg was selling at TSh 1,500 ($3) in Arusha. Transportation charges for both would be TSh 300 ($0.60).

In 1985, SIDO and ITDG established a small-scale lime production unit in Oldonyosambu but this did not last long. It was noted that the lime produced was not actually used in Oldonyosambu village but transported to Arusha. Today only six houses are plastered with lime in the trading centre but none in the village.

Misasi

This is a trading centre in Mwanza Region located about 70 km from Mwanza town. Accessibility to Misasi from Mwanza town is reasonably good. Roads from Misasi to neighbouring villages are not very good, making industrially produced building materials expensive and unaccessible. A 50 kg bag of cement costs TSh 4000-4500 ($8-9) in Misasi. Most houses are of mud blocks both in the village and in the trading centre, but well plastered in the centre. The types of binders used in the trading centre are cement and lime but a very small percentage (less than 2% of all houses) used lime as binder in the neighbouring villages.

One of those villages is Shilalo, where some limestone and coloured stones are found. These two types of stones are quarried and processed as lime powder and colour pigments respectively, for wall rendering. A group known as Athuman Pozzolanas is dealing with the small-scale production of these materials. The pigments from stone may be found in red, pink, green and yellow. These are widely used in the vicinity. This is a good area for further exploration as it has got ample potential both in limestone and coloured stones.

Kinyerezi

Kinyerezi is a village located in Ilala District, 25 km from Dar es Salaam. With bad road access, supply of building materials is difficult. The Kinyerezi ward has got a population of 3,500 people with a total of 200 houses, of which 120 are of mud and pole (60%). Where binders have been used, this has mainly been cement. Only very few houses (3%) have used lime, not as a binder, but as whitewash. The reasons forcing the majority of Kinyerezi people to build houses without binders are a lack of resources and the bad access road; therefore binders are very expensive. During rainy seasons it is not very easy to reach this place. There is no public transport to this village; people have to walk a distance of about 7 km from the main road or use private bicycles to reach Kinyerezi village. One bag of cement of 50kg costs TSh 3,000 ($6) where a 5kg bag of lime costs TSh 300 ($0.60).

Chinangali - Dodoma

Dodoma is the Capital City of Tanzania located within the Central Region of the country. Chinangali is located 5kms from the town centre. It is an unplanned area which was purposely blocked from further expansion by Nkuhungu low-cost housing project. The residents of this area have low incomes, and recently rushed to establish this new unplanned settlement in an attempt to free themselves from CDA (Capital Development Authority) building standards which are too high to meet.

The majority of houses in this area are small in size and are constructed using mud blocks. About 72% of them have cement floors. For plaster, about 40% used cement and 40% used lime, the remaining used clay soil or oil residue to protect their wall from erosion by rain.

The problems preventing residents of this area

from getting access to binders are their high cost and bad accessibility to this area. A 50 kg bag of cement costs about Tsh 3,500 ($7) and lime about Tsh 1,500 ($3). The means of transport is mainly by push-cart because the load in most cases is unreasonably small for motor vehicles which charge exorbitant prices.

Houses in which binders have been used
Materials selection

The use of about eight types of binders was noticed during this survey. These include cement, lime, pozzolana, cowdung, oil waste, ash, liquor residue and clay soil. Most of the non-conventional binders are locally available and were mostly used as temporary alternatives while gathering resources for using conventional binders, either cement or lime, although the majority ended up at this stage only. Each binder is used for a different purpose depending on the problem which the user is trying to solve.

Generally, about 55% of the houses are not more than 10 years old. Kigamboni is having more houses which are over 20 years old because the settlement itself is old and the houses were built using old technology and standards. These houses have thick walls and good foundation and lime is produced within the settlement.

The selection of materials depends mainly on:
- availability within the location
- affordability
- accessibility of the location
- technology and skills.

Foundations

About 30% of the houses visited used stones for their foundation. The areas reflecting substantial use of stones have abundant supplies of stones as compared to the rest. This is a clear indication that materials selection for foundations reflects availability and affordability.

Sand cement blocks have been used by about 19% of the houses, the majority being in Morogoro and Kigamboni where cement is much less expensive than in more remote locations. The use of sand cement blocks also depends on the availability of sand within the area to avoid transport costs. Most low-income earners tend to use the most economical mix of sand cement rather than the strength of the block itself.

The use of concrete for foundations is not a common practice for low-income people because all materials for making concrete: sand, cement and coarse aggregates are unaffordable to the majority. There is a wider use of concrete for foundations in Morogoro.

Mud block foundations are more popular in Dodoma because it is the cheapest material and technology available. The traditional house of the region, the *Tembe*, uses mud in situ walls, therefore mud blocks are just an improved version of their technology. The soil is also good for making mud blocks. Overall, about 15% have used mud blocks for foundations.

The survey found that 23% of the houses surveyed had no foundation. Arusha and Kinyerezi are leading, and this is mainly due to the type of wall material of their traditional houses, which is mud and pole.

Foundations of most of the house are faced with the problem of erosion around the house. Most foundations are not raised sufficiently and there is no proper and durable drainage around the house. The erosion problem is also magnified by rainwater from the roof especially where the roof overhang is small as is the case in many houses.

Floors

It was evident that, where stones were available like in Kigamboni and Mwanza, crushed stones are used as hardcore. Other areas used crushed rubble, compacted earth and sand. The preparation of the floor base is very critical for the durability of the floor finish to be laid on top.

Of the houses visited, 8% have no floor finish. House construction by low-income people is a long process, especially when using conventional materials like cement. Therefore owning a house without a floor finish is considered to be a transitional stage while mobilizing resources for finishing with other binders, preferably cement.

Problems associated with compacted earth floors include:
- the floor is dusty and needs frequent watering especially when sweeping the floor as it harbours insects,
- termites can easily pass through from below and destroy materials in the house.
- erosion especially when sweeping, particles of sand can be swept away.

Some low-income people may not be able to escape from the transitional or temporary stage because it is all about the ability to mobilize resources. To make life easier during this stage, there exist several surface hardening treatment methods using clay soil, cow dung and pozzolana. About 6% of the sample, mainly in Arusha and Mwanza and a few cases in Kinyerezi, used clay soil as floor finish on top of compacted earth floors. Clay soil has the capability to hold water for longer periods, therefore reducing watering frequency. The particles of clay soil are not as loose as compacted earth and therefore there is less erosion by sweeping. In an attempt to reduce erosion and dust of compacted floors, cowdung was found to have been used in about 8% of the houses. All those are in Arusha region where the people are pastoralist. About 8.7%

of the houses use pozzolana as floor finish while 2% used lime. Pozzolana and lime finish was only found in Arusha.

The majority of the houses, 83%, used sand cement screed as floor finish. It is the ultimate goal of the majority of respondents to have their floor finished with a sand cement screed. Respondents commented that floors with a cement screed are:
- easy to clean even with water
- durable
- modern
- hard surface.

As in the case of sand cement blocks, cement is used in the most economical mix, therefore one cannot expect durable and hard floors as required.

Walls

There are seven types of walling materials which were found to have been used in the six locations visited. These are sand cement blocks, burnt bricks, mud blocks, mud and pole, stones, pozzolana blocks and pozzolime blocks. Nearly half of the houses surveyed have used sand cement blocks for walls. Other types of walling materials were referred to as temporary and whenever improvements are done, most preferred sand cement blocks because they are more durable. But quality is really compromised due to the cost of cement.

Of the houses visited, 0.7% used burnt bricks for walling. This small percentage is due to a lack of fuel on the outskirts of most towns, therefore local production is not feasible and industrially produced burnt bricks are too expensive for low-income people.

Mud has been used as walling material by many low-income people. The material is easily available, the technology is simple and is used widely in traditional housing. Mud is popular in areas where soil is good for making mud blocks. 28% of the houses used mud blocks for walling. Mud is also used together with poles (wattles) to construct walls especially in areas where poles (wattles) are available. Overall, about 23.3% of the houses visited had mud and pole walls.

About 4% of houses surveyed had stone walls. The percentage is small because stone needs a binding mortar which in most cases low-income earners cannot afford. Pozzolime blocks have been used by 2.7% of houses all found in Arusha. The technology seems to be unknown to many low-income people. It was revealed that houses with this type of blocks are those made by institutions like the Building Research Unit, including a demonstration house, a dispensary ward and other public buildings in the area.

Roofing

The roof is another component of the house in which low-income people have few low cost alternatives. The conventional materials, corrugated iron (CIS) or aluminium sheets, tiles or asbestos-cement sheets are too expensive for low-income people. Apart from the high cost of CIS, about 88% of the houses surveyed used CIS for roofing because it protects the mud wall from erosion and getting wet. Another 3% used flattened tins as roofing, but these have a lot of disadvantages.

Most traditional houses with the exemption of *Tembe* in Dodoma utilized grass as the roofing material. About 10% of the houses in the survey used grass for roofing. In coastal areas they used palm leaves. The disadvantages associated with the use of grass include:
- insect/termite attack
- fire hazard
- frequent replacement
- grass becoming scarce and costly.

Materials used for plastering and rendering

Tanzania's low-income people, as those in most developing countries, are using less durable walling materials and if left uncovered/unplastered, the house would not be as beautiful. They therefore have a number of alternative plastering and rendering materials for either permanent or temporary use while gathering resources for more superior binders. Materials used for plastering most of the houses visited have been behaving differently depending on the type of walling materials used.

About 56% of the houses used sand cement plaster. As stated earlier, this type of plaster has good binding strength with sand cement block walls, but does not represent the majority of the low-income people. Sand cement plasters on mud walls (mud blocks or mud and pole walls) have been performing very badly. The problem has been lack of bondage between the two materials i.e. plaster and mud wall. The plaster tends to bind to the outermost particles of mud walls which, with the differences in temperature experienced especially on external plasters, leads to the cement plaster detaching itself, resulting in a gap between the two. This makes the plaster more vulnerable to cracking and breaking when knocked by hard objects. Despite the bad behaviour of sand cement plaster, most respondents still prefer to use it because it completely changes the look of the house. Cement plasters are more practiced in locations near towns with a cement factory.

The use of lime for plastering was noticed in about 27% of the houses interviewed. Those who used sand cement blocks on walls preferred to use cement plaster, therefore the majority who used lime plasters had their wall made from mud. There were less problems of cavitation between wall and plaster when lime was used. This is due to smaller differences in thermal expansion between mud walls and lime plasters. However, in trying to eliminate these problems, some respondents used soil lime mortar for laying the mud blocks. However soil lime plasters are not as hard as soil cement plasters, therefore they can easily be notched by hard sharp objects.

Mud plaster is only applied on mud walls. The main purpose is to protect the weak wall by providing a new layer. Mud walls and plaster are more vulnerable to erosion by wind and rain water from driving rains or from roofs especially when the roof overhang is small as is mostly the case. Abrasion by users and animals also contributes to erosion. To protect mud wall and plaster, low-income people are applying different surface rendering methods depending on the availability of materials within the location. These methods include rendering with cowdung which was the case for about 8% of the houses, oil waste painting for about 4% and conventional paints for about 1%. Other methods observed include rendering with liquor residues and ash from burnt coconut trees.

About 10% of houses visited used pozzolana for plastering. The material is more durable than mud plastering and is practiced more in areas where it is available like Arusha. Mbeya region, although not visited, is very rich in pozzolana.

Clay soil plaster is not much different from mud plasters. It is as vulnerable to damage and bad weather as mud plaster. Surface treatment methods are also applied to this type of plaster to protect it from being exposed. Generally about 13.3% of the houses were found to have been plastered with clay soil.

Construction and maintenance

As mentioned earlier the construction of houses for low-income people is more of a process than an activity. In most cases house owners involve themselves in the real construction because the technology has been passed to them by their elders. Men usually do the real construction and women participate in such activities as fetching water or building materials and rendering walls. Where the house owner lacks the skills, especially in the use of conventional binders like cement and lime, they hire artisans to do the job. To cut down expenditure on labour charges, house owners tend to exhaust all human power available in the house. Community participation has never been the trend in all locations visited.

Soil and organic materials, which most low-income people use for the construction of their houses are not durable. They get damaged and therefore are replaced almost after every rainy season. Floorscreeds and plasters which provide protection to less durable materials had a lower frequency of repair. About 35% of respondents had either repaired the damage or replaced it for the purpose of improvement in the last six months. Wall renders with cowdung, oil and liquor residue are frequently repaired or replaced after every rainy season. This job is mostly done by women. 50% of the respondents had not yet repaired or changed their plaster or floorscreed because they were still new.

Problems associated with availability of binders

Local or unconventional binders like cowdung, claysoil, pozzolana etc. have no problems as far as availability is concerned. Most of these alternative local binders are available at the locations, and solely used where they are locally available. The problem is purely in their properties which make them to be referred to as temporary.

Conventional binders i.e cement and lime are produced in a few areas of Tanzania. Their availability in the survey locations has been less of a problem than poor access, poor transport and a high price. In the absence of building societies and community owned shops, the supply of lime and cement has been left to private individuals who purely supply at competitive costs. Therefore, because there is less use of these materials in low-income areas, it means there is less demand and therefore little or no supply. 9.3% of respondents feel that availability is the problem. Most respondents discarded the problem of availability and instead argued that it is the question of affordability of materials.

The cement industry in Tanzania is owned by the state. In recent years Tanzania has been undergoing an intensive Economic Reform Programme monitored by the World Bank, and has therefore been less sympathetic on subsidizing essential commodities such as building materials like cement. The price of cement has been rising steadily due to the increased production and fuel costs. 60% of respondents complained of higher prices of cement and lime.

Lime is produced by both the public and the private sector. The private sector is purely operating on a commercial basis and prices are at market value. Local production of lime in areas where limestone is available reduces the magnitude of problems within the location.

About 38% of respondents find that transport costs are higher. This is contributed by:
- bad accessibility
- distance to where the binders are purchased
- the rise in fuel and maintenance costs.

Transportation in such locations is done by vehicles, bicycles and push carts.

About 15.3% of respondents found that labour charges are higher. Artisans had a tendency of thinking that when someone wants to use conventional binders they must have higher incomes and thus pay more than others.

Quality

About 72% of the respondents used binders because of the following reasons:

- durability
- prestige
- health i.e. dust, insects, termites, bugs
- maintenance: easy to clean and no frequent repairs
- beauty
- modern aspect

The remaining 28% had not used binders because the houses were temporary.

The survey revealed that 98% of respondents were willing to be educated about the use of new types of binders with the expectation that they might be cheaper and more durable. About 37% were satisfied by the quality of their houses. The remaining 63% said it is a question of affordability but almost everybody must aim for a much better life and house.

About 44% of the houses did not have a good appearance. It seemed the majority were not innovative in designing their houses; the construction revealed poor skills and workmanship. Most of the house designs were very simple just to cater for basic functions and to accommodate the family. About 24% of the houses observed had inadequate space. About 28% of the houses had unstable walls and 30% had poor or unstable floors. Leaking roofs affected about 27% of the houses due to poor covering and roof skeleton structures.

About 40% of the houses had unreasonable windows sizes. The materials used for windows and door frames were poles and timber; for panels tins, CIS, boards, and timber are used. Internal doors can or may not have panels at all. Drainage around the houses had taken its toll on about 42% of the foundations of the houses; repairs on foundations had to be effected after or even during rainy season.

Quite a number of useful indigenous technologies were noticed which could be promoted in addition to the more common binders cement and lime. The artisans (fundis) encountered had worked mostly with cement (63%), followed by lime (20%).

The survey revealed that binders selected are cement, lime, clay soil, pozzolana, ash, cowdung and liquor residues. Cement is most often used because of the following reasons provided by the fundis:

- superior quality
- easier available than lime
- easier to work with than lime
- local builders are more conversant with it than with any other binder
- low level of awareness of alternative building materials amongst builders as well as local communities.

Houses in which binders have not been used

Housing conditions

The slogan 'A bad solution for tomorrow is a good one for today' remains true for low-income communities. This is justified by the use of non-conventional binders in rural areas where they use materials which are within their reach and literally free of charge. These materials include clay soil, cowdung, liquor residues, waste oils, cassava and ash. Respondents in the household budget survey of 1991/92 showed a number of people saying they do not need to improve their houses. For them, an improved house of today is better than an improved one of tomorrow. Assuming their reason is low economic power, if one adds the percentage of those with no money, and those with no time, and other reasons the percentage for not improving house comes to 89%. This is alarming for housing professionals who must be asking themselves how the government, NGOs and professionals effectively can improve low-income housing.

Looking into the conditions of houses without binders in comparison with similar houses, of the same age, for which binders have been used, one can easily see the importance of using binders. Most of the houses in which binders have not been used are in pathetic condition. On the basis of the findings of this research, it is recommended that individual families which cannot afford binders be assisted by introducing, and encouraging the use of alternative binders like lime, which is far cheaper than Portland cement. The use of alternative binders in providing shelter for such people would raise the standard of their accommodation, hence their lives.

The researchers surveyed 60 houses in total. The surveyed houses had different life spans. The majority, 45%, had a life span of less than 6 years; 21.6% of 6-10 years, 5% of 11-15 years; 8.3% of 16-20 years and 8.3% of above 25 years.

Materials

Most of the dwellers interviewed demonstrated that they were heavily dependent on construction materials used in their natural rudimentary processed form. They had simple traditional structures built of non-durable and locally available materials. The houses visited also varied according to cultural preferences and climatic conditions. This resulted in a wide variety of layouts, designs and materials used. For example some houses were circular in shape while others were rectangular.

Foundations

The study revealed that where mud and pole was used as walling material, none of the houses did have foundations. Small holes of a depth of 50-60cm (an arm length depth) were excavated for each vertical pole, the poles would then be fixed in the holes and compacted with soil before the rest of the walls are erected.

In the cases where compressed mud blocks technology is used 36.7% had foundation made of the same compressed mud blocks. Usually 30cm of the top soil is excavated so that the mud blocks are laid directly on the firm soil. In the surveyed areas where stones are plentiful, 30% of the houses visited had stone foundations.

Floors
Almost all dwellings visited had tamped soil as floor (93%). The remaining 7% had floors made of compacted cow dung.

Walls
The most common wall materials in the survey of houses where binders have not been used is mud and pole (43%). Mud and poles are arranged such that a rectangular network of poles is built. One line of vertical poles is tightly built in the centre of the wall and horizontal lines of reeds on each side of the vertical poles. The rectangular opening between the poles and reeds are then filled with soil. Proportionally there is more soil in these walls than poles and reeds. There is another type of the same walling materials which prefers the use of more poles than mud. In this method vertical poles are placed close together and the horizontal reeds are few and basically used only to keep the vertical poles together. The soil, sometimes mixed with ash or cowdung, is used to fill in the small gaps between the vertical poles. This type of wall formation is largely found in the dusty/silty soil areas of Northern Tanzania where thick poles or split poles are used. It is also found in predominantly clay soil areas where thinner sticks are used as basic wall structures in combination with soil/mud in the same way as above.

The second most common group of wall materials found in this study is compacted mud blocks which constituted 40% of houses surveyed by the researcher. The in situ soil walls constituted 13% and stone walls 3%.

Roofs
The most common roofing materials used is grass. 47% of the surveyed houses used grass. The use of palm tree leaves is popular along the coast of Tanzania and grass in other parts of the country. 33% of households used corrugated iron sheets. Flattened tins accounted for 3% of roofing materials.

Wood poles and reeds are the most common ceiling materials. Normally soil with or without cowdung is used as infill between the poles or reeds. In coastal areas small pieces of coral stones are applied as ceiling materials. All the ceilings observed were flat.

Construction
The study revealed that 42% of the houses in this category were built by male house owners; 30% by contractors/artisans and 25% by both men and women members of the household; 3% were built by women headed households. A large majority (72%) of the respondents commented that community participation is not practised while 28% responded in the affirmative.

Constraints for using binders
The study revealed that 80% of the households visited did not use binders because they are expensive, thus unaffordable. 15% appeared to be ignorant of the technology. 5% gave other reasons as a major constraint; they included non availability and transport problems from the purchasing places to their communities.

Quality
The majority of the surveyed households were of the view that their houses were of poor quality, 77% of the people thought their houses were not up to standard and only 23% were satisfied with the state of their houses. Those who expressed satisfaction argued that they could not think of better alternatives due to poverty and lack of financial resources. On the other hand households which expressed dissatisfaction identified the following problems associated with their houses:

- the mud infill can easily be removed or damaged because there is no strong bondage between poles and mud especially where the wall is thin.
- most floors are dusty, therefore making it easy for insects to hide and difficult to clean.
- flooding during rainy season, because sometimes the floor level is below that of the outside ground.
- poles can be damaged by insects and termites
- walls can be eroded by wind and abrasion by users.
- if not protected they can easily be eroded by driving rain and from the roof if the roof overhang is small as is the case in most locations.
- repairs have to be done after every rainy season. The study revealed that 42% of the dwellings were repaired in the last six months while 58% were not repaired although not in good shape.
- openings are small, ventilation is inadequate; it is dark inside, therefore unhealthy to live in.
- walls create places for insects and other pests to hide and are difficult to clean.
- leakage of the roof.

House owners interviewed have the following priorities for further development:
- 42% would wish to demolish the existing houses and build new ones.

- 15% have no priority at all.
- 12% would wish to improve the walls of their houses by plastering and put cement screed on their floors.
- 3% would wish to extend the existing building.

When the heads of households were asked whether they are ready to adopt the use of binders, 97% responded positively for the following reasons:

- strength
- durability
- ease of cleaning
- good appearance
- health

There were 4% who responded that they were not ready to use binders because they are already too old to be engaged in changes where financial resources are a constraint.

Most of the houses did not have a good appearance; 62% were very ugly while 39% appeared reasonable. The study revealed that 57% of all the houses observed had adequate accommodation for sleeping, resting and cooking while 43% had inadequate space. The research findings showed furthermore that 69% of the houses surveyed had unstable wall structures; the reasons for this are the lack of proper foundations and the wearing out of the mud walls during the rainy season. 33% of these houses had poor roof structures leading to the collapse of the roofs in pieces; 33% of the houses observed were leaking.

Applied skills and technology
Cement
Most of the artisans (fundis) were familiar with the use of cement, but, as mentioned before, mixing ratios were not strictly followed due to price implications. The mixture depended a lot on the artisan and the owner of the house. Normally they get ratios by volumes, using the pail or *karai*, e.g. 1 karai cement:6 karai sand for floor screeds.

They also have a fair knowledge about other important factors in the use of this binder, e.g.:

- sand should be free from other particles like stones, grasses, timber pieces etc.
- use of a straight edge for plastering and levelling the floor.
- wall rendering and floor screed is cured for three to seven days.

Lime
Very few builders could give clear technical details on lime applications. It was learnt that lime is most commonly applied as a wall paint or render, but not for masonry, plasters or floor screeds. Some artisans think that lime is more expensive in use than cement, because it needs richer mixes to achieve a similar quality.

Pozzolime
Pozzolana alone is not a good binder. It needs activation by either cement or lime. Fundis in Oldonyosambu and Arusha, who have worked with the lime-pozzolana produced some years ago in that region, explained that mixtures of 1 lime: 3 pozzolana: 3 sand do make good masonry mortars and plasters, and mixtures of 1 lime: 3 pozzolana: 3 pumice have been succesfully used to make concrete blocks or light concrete.

Ash and others
Due to the fact that these binders are not so common, the technologies involved were not explored fully. But a mixture of ash and cowdung has proved to be very successful for plastering both in Arusha and Mwanza.

The survey revealed that local builders would like to be assisted technically on floors, walls and plasters. The type of binders to be promoted are lime, for all locations, but mixed with pozzolana for Arusha. In a very special way, builders in Mwanza would like to promote soil, liquor residues, and cowdung as appropriate binders because they are locally available and affordable.

Conclusion and recommendations
Affordable shelter of reasonable quality cannot be achieved without proper skills and technologies, especially when one is concerned with low-income people. The speed of increase in building costs is obviously unproportional to the income growth of low-income households, making conventional building materials unaffordable to low-income earners. Therefore, there is a need of promoting alternative building materials and technologies which may be affordable by the those groups.

The promotion of alternative binders cannot be generalized for all the six locations. Lime is recommended to be promoted for use by low-income people in all locations, but the locally available binders should be given first priority.

The survey conducted realized that houses without binders are generally in pathetic condition. The best way to redress this situation is to make use of alternative binders which may be available in the country, and can be produced locally, such as lime and pozzolana. It needs to be emphasized that alternative binders like lime and pozzolana have not found wide scale use because of user prejudices. Similarly, a variety of earth based technologies are available for combination with lime, e.g. lime-stabilizing earth blocks with lime are cheaper and durable, and lime mortar could be used for laying mud blocks. After years of research these technologies are only applied to a few demonstration houses with a minimum impact to intended beneficiaries.

As observed above, the use of alternative binders

is a move towards creating access to housing at an affordable cost. Lack of appropriate standards for such materials is a stumbling block to their wide scale adoption. In the absence of such standards the materials are seldom included in specifications for houses built in urban areas. Another barrier to increased utilization of alternative building materials in shelter production is the inflexibility of existing building regulations. The government frequently exercises considerable indirect control over the development of building materials through building regulations, codes and standards. In rural areas there is no handicap on the use of innovative materials because standards do not apply there. There is still a need to promote the use of alternative binders, but professionals and researchers have to convince the government of their potential. With regard to the present study, people have expressed a willingness to use alternative binders and existing prices seem to be affordable by low-income earners. Therefore the future market of alternative binders could be bright.

More often, professionals tend to ignore local skills and technologies. The study has shown that there is a lot of potential in using alternative building materials, most of which are not yet utilized to their full potential, particularly in rural areas. To achieve affordable and durable shelter, the potential of alternative binders must be firmly established and explored. Moreover, most of the artisans interviewed aspire to develop alternative building materials and technologies especially for innovative binders and are ready to apply them in house construction.

There is a great need of creating an awareness on the use of alternative binders like lime instead of Portland cement. Related advantages and disadvantages and their cost effectiveness should be well addressed to people. This will offer an interesting model for the promotion of alternative binders. This could be done by research institutions like the Centre for Housing Studies, Ardhi Institute, through radio and television programmes, newspapers and publication of small brochures in simple local language, i.e. Swahili. Training should be provided to local artisans, *fundis,* on the use of alternative binders, through short courses. And some further research may be needed, e.g. on plastering of earth constructions.

The government should influence desirable changes in regulations, codes and standards that currently have the effect of inhibiting local building materials and the introduction of innovative technologies. Such changes should be localized and flexible. The use of alternative building materials and technologies should be fully adopted in urban development projects.

The government should facilitate the expansion of the domestic production of alternative binders by making raw materials easily accessible to producers and by providing support to investors, either by acquiring selected technologies or by improving the efficiency of existing technologies.

There is a need to establish building societies which will be in a position to provide building finance as mortgage or supply materials to builders. This system may also be structured to make building materials available in building localities.

References:

Planning Commission: "Economic Survey", Dar es Salaam, 1994

A.L. Mtui and G.M. Kawishe: "Portland Pozzolana Cement", Dar es Salaam, 1983

Industrial Communities: "Quarterly Report", June 1994, Dar es Salaam

Yrjo Tolonen: "The Consumption of Lime in Residential Buildings", Dec. 1983

Wilt de Boer: "Appropriate Building Materials and Constructions, Parts I& II", Dar es Salaam, 1983

Bureau of Statistics: "Household Budget Survey, 1991/92", Dar es Salaam, 1994

Zanzibar

Theo Schilderman

Historical uses of lime
Zanzibar is one of the earliest settlements in East Africa; as a consequence it has a very rich history of using lime which started in the 18th century. Lime was the predominant binder in the historic Stone Town of Zanzibar for at least two centuries; many old houses of several stories height are still standing to prove that. Outside the Stone Town, where most construction is low-income housing, there is a more limited, but still important use of lime, mostly for wall construction and whitewash.

Lime is traditionally produced by heap burning, mainly using coconut timber as fuel; there have been a few attempts at using permanent vertical lime kilns, but apart from the one at Dunga which is still irregularly used, this production method has not really taken off.

Nowadays, lime is mostly used in its powder form (lime hydrate), though historically the use of lime putty was common; because higher quality can be achieved with lime putty, this is again being revived by the Stone Town Conservation and Development Agency (STCDA) in its conservation work.

Foundations and walls
The historic houses of the Stone Town have massive foundations and walls of ragstone, held together by lime mortar; for this mortar, the lime was generally mixed with a lateritic type of soil and sand, in varying proportions. In low-income housing, a similar type of mortar is sometimes used as infill in mud-and-pole type walls.

Floors
The historic houses, which had several floors, used mangrove poles to support their floor slabs, made of a mixture of lime and mostly laterite, which was very well mixed and compacted, to make the floors watertight. The floor screeds used a lime-sand mix.

In rural areas, floor screeds are known to have been made from a mixture of lime, clay, sand and cow or donkey dung, whereby the fibres of the dung helped to prevent surface cracking and to improve watertightness.

Plasters
Most traditional plasters have three coats; the first two are made of a mixture of lime, sand and laterite, often in a proportion of 1 lime : 2-3 aggregates, whereas the skim coat would be made of pure lime putty.

When lime plasters are used in modern construction, e.g. on concrete blocks walls, a first layer of cement-lime-sand plaster is sometimes applied, followed by a layer of lime plaster.

Decorative stuccos
These are one of the most important features of the historic buildings of Zanzibar and show the most intricate use of lime; these stuccos are of enormous historic importance. Stucco work is an embellishment, trying to imitate for instance stonework or carved timber; it was achieved by using running moulds, cast moulds or hand modelling. Zanzibar has many fine examples of lime-stucco modelled in-situ, carved and moulded rose windows, moulded column bases or architraves, and many other elements. Various additives were used in lime stucco, some of which were organic oil-based, others pozzolanic, e.g. brick dust.

Due to lack of maintenance, many buildings in Zanzibar are on the point of collapse, and several have been lost in recent years; once this happens, the historic stucco is lost forever.

Whitewashes
Most houses in the Stone Town were originally whitewashed in two or three coats, and this practice also occurs elsewhere, on a moderate scale. When using a pure lime-water mix, such whitewashes do not last very long. Various additions are known to have been used, including coppersulphate acting as a fungicide, and boiled seaweed or sugar to increase its abrasion resistance.

Current uses of lime
The use of lime is on the decline in Zanzibar; its current use in building is mainly restricted to general mortars and plasters as well as whitewashes, with the exception of some minor other uses in conservation. Yet, the use of lime in Zanzibar is still the most widespread in the region as, for instance, the figures in Section 1 indicate. Factors that have contributed to the decline in the use of lime include:

- preference for cement which had certain advantages over lime such as faster strength development and maybe has a higher status,
- decline in the quality of lime caused by the shift from lime putty to lime hydrate,
- loss of skilled craftsmen, particularly in stucco work.

Currently, lime is often considered a second choice after cement; unless the quality of the product and the skills involved are improved, this will probably remain so.

In practice, cement has not always been the best material. It has often been used in repairing old buildings, and in the process caused more damage than it resolved. The application of a watertight cement plaster to an old wall has the effect of trapping the humidity inside (whereas a lime plaster would allow the wall to breathe), causing structural damage to the wall, and making the humidity appear elsewhere, complete with salt efflorescence.

The conservation of past technologies

As mentioned before, Zanzibar's Stone Town is under serious threat. A combination of factors, such as the departure of the previous rich merchant families which had built the historic houses, as well as the gradual disappearance of skilled craftsmen, has caused a serious lack of maintenance of most of the old buildings. Leaking roofs and pipes, capillary water, incompatible cement renders and other factors have already done a lot of damage; each year, several buildings are collapsing, and the list of buildings under threat is growing.

The STCDA, with the help of outside agencies such as UNESCO, UNCHS or the Aga Khan Foundation, is making a great effort in trying to conserve the great historic, cultural and architectural heritage of Zanzibar's Stone Town. In the past decade or more, extensive surveys have been done which have allowed a greater understanding of Zanzibar's building traditions. Several buildings have been restored, and whilst doing so, old skills have been revived. The STCDA now has a core body of architects, technicians and artisans with a substantial understanding and experience of traditional lime technologies. ITDG has assisted with the reintroduction of the lime putty technology and some of the traditional plasters including the use of additives such as pozzolanas. But a lot more work is still needed, for instance in improving lime production itself, e.g. by the introduction of intermittent kilns which can produce a higher quality lime.

Within the East African region, Zanzibar remains the centre of excellence for the application of lime. This is an invaluable asset which should be further exploited, e.g. by allowing technicians or craftsmen from elsewhere to be trained in Zanzibar. On the other hand, there have been developments in lime production elsewhere in the region, of which Zanzibar could benefit. A two-way regional technology transfer seems therefore desirable.

References:

Stafford Holmes: "Advice on the use of lime in traditional building, Zanzibar, May 31-June 16, 1989", ITDG, UK

Stafford Holmes: "Advice to the Stone Conservation and Development Authority on lime technology and historic building conservation, November 17 - December 1, 1990", ITDG, UK

Stafford Holmes and Michael Wingate: "Report and advice given to the STCDA for emergency repairs, traditional lime technology and small-scale lime production, November 16 - December 2, 1991", ITDG, UK.

Fatma I. Kara: "Traditional and current uses of lime mortar, render and stucco in Zanzibar" in: N. Hill, S. Holmes and D. Mather: *Lime and other alternative cements*, ITPublications, London, 1992.

Archie Walls: "The Revitalization of Zanzibar Stone Town" in: Abdul Sheriff (ed.): *The History & Conservation of Zanzibar Stone Town*, Dept. of Archives, Museums & Antiquities, Zanzibar, in association with James Currey, London, 1995.

SECTION 5

Workshop resolutions and recommendations

Resolutions

The Workshop resolved that:

(a) Energy efficient kilns be designed, developed and promoted to cut down on the rate of fuel consumption.

(b) In addition to mandatory afforestation, sustainable use of fuelwood by lime producers must be ensured.

(c) Geological Departments together with Lime Institutions should map out and evaluate all the various raw materials for alternative binder production.

(d) Provision be put in place for training of Artisans, Professionals and Entrepreneurs in the use of Alternative Binders.

(e) Standards and Codes of Practice for the quality and application of the different binders be established.

(f) Relevant information on the binders be disseminated e.g. through networking, exchange visits, etc.

(g) Further research be carried out with respect to improved kilns; the applications of alternative binders; and a more intensive use of waste as fuel and pozzolana,

(h) In recognition of the capacity of the Intermediate Technology Development Group in terms of their extensive networks, long experience and sustained interests, ITDG be mandated to work in collaboration with other interested agencies in the further exploration and implementation of the resolutions of the Workshop.

To achieve the above, the following recommendations and actions are proposed to be undertaken:

Problems and Recommendations

1. Raw materials

Potential producers have inadequate information about the location of raw material deposits and their available quantities and qualities. And in the case of some specific binders, e.g. pozzolanas, they may even be unaware that certain materials in their vicinity could be used as such. In some cases, this information is simply not available, but when it is, it is not disseminated to small producers. Once sources of raw materials have been identified, they face a further problem that they may not be of homogeneous quality, and that simple methods of quality control are lacking.

Another problem is that taxes and levies on raw materials may be quite high.

The Workshop therefore recommends that:

1.1. Limestone, pozzolana and other raw material deposits be mapped, quantified and qualified.

1.2. This information be disseminated to existing and potential producers.

1.3. Producers get raw materials analyzed before embarking upon production.

1.4. Producers be encouraged to form associations in order to facilitate access to information about and testing of raw materials.

1.5. Appropriate testing methods be developed for use of small producers.

1.6. Extension services be provided to producers on raw materials identification and testing.

1.7. Taxes and levies on raw materials should be reduced.

2. Quarrying

Quarrying of limestone is often being carried out by blasting; the use of blasting equipment and explosives is expensive. Quarrying by small producers is also often done in an unplanned and haphazard way, and there is frequently a lot of quarry waste. Access to some quarry sites is difficult. And finally, quarrying is subject to regulations, licenses, and lease arrangements which may be time-consuming to deal with, often of short term duration, and costly.

The Workshop therefore recommends that:

2.1. The nature of raw materials should determine whether blasting is necessary.

2.2. Where possible, quarries should be located close to access roads and to production sites, but where this is not possible, access roads should be provided.

2.3. Extension services be provided to producers on more systematic quarrying.

2.4. By-products such as aggregates, animal feeds or agricultural lime be made from quarry waste.

2.5. Governments should review favourably existing laws, regulations and taxes pertaining to quarrying.

3. Size and type of production

These are to some extent determined by the available quantity and quality of raw materials, and by potential markets for the products. Small production units are easier to manage and to start, and require less capital; they also generate relatively more employment. But they do not benefit from economies of scale and may be inefficient in the use of fuel. Bigger production units are able to satisfy large orders, capable to produce a higher quality product, and may offer a higher return on investment. But there is also a lot more risk involved in big plants, of capital lying idle when there is no demand, and of workers being laid off.

The Workshop recommends that:

3.1. The size of production units should be determined by the market in the first place.

3.2. Labour-based production methods should be promoted to generate employment.

3.3. Small producers are encouraged to form associations or co-operatives in order to benefit from some of the advantages of larger scale production, e.g. access to improved technologies, investment capital etc.

3.4. Vertical shaft kilns be used for lime production.

3.5. Small-size ball mills and, where necessary, driers be used for pozzolana production.

4. Production technology

In the case of lime production, one of the main bottlenecks is the cost of a kiln which may be excessive for a small producer. One of the major items contributing to that high cost is the refractory bricks used to line the kilns.

Traditional production uses a lot of fuel, partly because kilns are not well insulated. Small producers also have insufficient information about the various kiln options. Feeding the kiln with limestone is frequently done by trial and error, without knowing the appropriate size of kiln feed; and limestone is reduced to size manually. To achieve a sufficiently fine end product, ordinary sieves are inadequate, but air separators are expensive. Finally, many producers are economizing on packaging by using unsuitable materials.

There is relatively little experience in the region with pozzolana production; most of this has only been on an experimental scale, and a lot of the technology remains to be further developed and tested in practice; this includes for instance rice husk burners, driers and ball mills.

As to the other alternative binders, such as gypsum, even less is known, and any improvement in their production or application would probably have to start with further research into their characteristics and qualities, in order to be able to define potential improvements.

The Workshop therefore recommends that:

4.1. A careful study is made of lime kiln options before a decision is made on kiln construction.

4.2. A catalogue of appropriate and proven lime kilns is elaborated and disseminated to producers.

4.3. Research is done into the local production of refractory bricks.

4.4. Alternatively, high quality burnt bricks, made from carefully selected clays with some refractory properties, could be used to line kilns.

4.5. All lime kilns be properly insulated; insulation bricks can easily be made locally without great expense.

4.6. Kilns are operated continuously to be more energy efficient; batch kilns should be minimized.

4.7. Extension services be provided to lime producers on the sizing of kiln feed.

4.8. Where funding permits, small crushers be installed for kiln feed sizing.

4.9. Where topographical conditions allow, plant layouts should take advantage of gravity.

4.10. A simple mechanical air separator for lime be developed and disseminated, using a generator whenever affordable.

4.11. Lime be properly packed and labelled, using good packaging materials.

4.12. Appropriate technologies for the production of pozzolanas be further developed or adapted, tested and disseminated.

4.13. Further research is done into the characteristics and improvements, application and dissemination of very low cost binders such as traditional stabilisers and special earths. In particular, research is required on pozzolana cowdung mixes, which show promising results near Arusha, in order to establish mixing ratios and other characteristics.

5. Marketing

Alternative binders are not very well known and need to be popularized; several of them are considered low status and most are not so easily accepted, particularly compared to Portland cement, which has recognizable advantages such as faster setting and higher strength. Some of these materials such as pozzolanas are innovative, and face the same marketing problems as other new products. The lack of awareness of alternative binders affects many actors, including users, artisans, professionals and policy makers. As to the practicalities of marketing, access and distribution are often problematic. But small producers also fail to advertise or demonstrate their products, and often lack general marketing skills. And sometimes their prices are too high due to high production costs or inefficient production.

The Workshop therefore recommends that:

5.1. Alternative binders be popularized using a variety of methods, including national shows, mass media, advertising, demonstration projects, use by producers on their own sites, seminars and training of users, artisans, professionals and policy makers, involvement of community based organizations, manuals and other publications, etc.

5.2. Extension services be provided to producers on marketing.

5.3. Producers be encouraged to form associations or co-operatives in order to reduce production costs and to enable joint marketing and influencing.

5.4. Pozzolanas deserve particular attention in the transfer of information and knowledge and should be popularized in their regions of occurrence.

5.5. The use of coloured pozzolanas, e.g. the Athuman pozzolanas in Mwanza region of Tanzania, is to be promoted.

6. Use of Alternative Binders

Some alternative binders, such as lime, special earths and cowdung are traditionally used in construction. But there is a tendency to replace them with cement, wherever affordable. This is partly due to their sometimes low status and perceived or real problems of quality. Small producers sometimes areunable to control the quality of their end products, whereas users may not always have the skills to apply them well.

The Workshop therefore recommends that:

6.1. Alternative binders are popularized, as mentioned under 5.1.6.2. Appropriate testing methods be developed for quality control by small producers.

6.3. Extension services be provided to small producers on quality control.

6.4. Training be provided to users and artisans on appropriate technologies of use.

6.5. Pozzolanas in particular are popularized and disseminated as mentioned under 5.4 and 5.5.

6.6. Binders other than lime and pozzolanas should be researched, further, made socially acceptable, and popularized in urban areas, taking advantage of relaxed bylaws.

7. Standards and Regulations

The existing standards, bylaws, codes and regulations pertaining to housing and building do not favour the use of alternative binders. They are restrictive and at an unnecessarily high and therefore unaffordable level. They clearly forbid the use of some of the binders listed above, and by not mentioning them, prohibit the use of others that might have the required qualities. On the other hand, a perceived lack of quality is very often behind the lack of popularity of certain binders, and their standardization might help their dissemination and sales.

The Workshop recommends that:

7.1. Existing standards etc. be amended to incorporate and facilitate the use of alternative binders.

7.2. There should be a Code of Practice specifying the quality and use of these materials.

7.3. The above be elaborated so that different standards may apply to different products for different uses.

8. Capital and Finance

Small producers need investment capital as well as working capital; the lack of it is a serious bottleneck for the introduction of improved technologies. Apart from that, such producers often also lack information about potential funding sources, are unable to produce bankable projects, lack collateral and credibility, and are faced with high interest rates. Their financial management is frequently weak, and they are subject to prohibitive taxation in various forms. A lot of these problems are not specific to the binder industry; they affect small enterprises in general. Besides, funds for research and development and technical assistance to producers and users are very limited.

The Workshop therefore recommends that:

8.1. Governments should favourably review the taxation and credit facilities for small enterprises producing binders.

8.2. Promotional Funds be established to provide capital to the small binder industry as well as supportive research and technical assistance.

8.3. Training and extension services be provided to small producers on financial management and the sourcing of funds.

8.4. Small producers be encouraged to form associations or co-operatives for easier access to assistance and information, credits and licenses and for more advantageous taxation.

9. Infrastructure and Transport

Road access is a common problem in the exploitation of raw materials. But the lack of adequate roads and transport facilities can also hamper the marketing and distribution of binders; since binders are dense, a decentralised market can considerably increase costs for the distant user. Finally, the absence or irregularity of electricity supply does pose constraints on production.

The Workshop recommends that:
9.1. The size of production units be established in relation to potential markets and particularly the costs of distribution.
9.2. Plant locations should be determined by the proximity of raw materials.
9.3. Governments be urged to improve transport infrastructure.

10. Energy and Environment

A major problem in current lime production is the excessive use of fuelwood, to a large extent due to very fuel-inefficient production methods, particularly heap burning and to a lesser extent batch firing. The use of fuelwood is unplanned and leads to deforestation.

The depletion of forests is an environmental concern. But there are also environmental problems related to quarrying, including the accumulation of waste, the unplanned use of deposits, the pits left open after extracting the raw

materials, and erosion and loss of biodiversity in coastal areas. Finally, there are problems of dust in production, especially during the slaking and bagging of lime.

The Workshop therefore recommends that:
10.1. Fuelwood should be used more efficiently via the dissemination of more fuel-efficient kilns, which are used continuously, and are properly insulated.
10.2. There should be compulsory afforestation by users of fuelwood.
10.3. Where other fuels are accessible, these should be exploited.
10.4. Where possible, the use of biomass should be explored as an alternative or additional fuel.
10.5. Where possible and acceptable, low-energy binders should be promoted, including for instance the partial substitution of lime or cement by pozzolanas.
10.6. The use of agricultural and industrial waste should be encouraged as sources of fuel and pozzolana.
10.7. Mining in coastal areas should be regulated.
10.8. Quarries and quarry waste should be used more systematically, as mentioned under 2.3 and 2.4., and reclaimed after use.
10.9. Producers should be informed and required to comply with existing rules and regulations; where they do not exist, these should be developed and popularised.
10.10. Dust extractors must be installed at the slaking site.
10.11. Workers must be provided with the correct protective gear, e.g. dust masks, goggles, overalls, boots, helmets and gloves.

11. Technology Transfer, Information and Training

Small producers lack information on technological alternatives which might improve their production. They also often lack know-how on various aspects of production and marketing and management.

Potential users are ill-informed about alternative binders; these are not included in the curricula of training institutions, nor is other information disseminated. The artisans in particular often stick to the materials they have experience with, e.g. cement, rather than venturing into the unknown.

The Workshop therefore recommends that:
11.1. Training needs in lime production be assessed and attended to via courses, seminars, workshops, etc.
11.2. Producers are encouraged to organise themselves in groups for easier exchange of information.
11.3. Exchange visits be organized.
11.4. Audio-visual documentation of production and use be prepared.
11.5. A manual on the proper use of alternative binders be produced.
11.6. Existing curricula be reviewed to incorporate alternative binders.
11.7. Networks be established between producers and users, at national and regional level.
11.8. Technical assistance should be funded by donors.
11.9. Information from outside the region should be accessed via subscriptions, publications and networking.
11.10 Budgets be provided for the above.

12. Research and Development

As mentioned before, some of the alternative binders are still relatively unknown, and still need verification which in turn requires further R&D. Similarly, production technologies can hardly be improved without this, and the same applies to the various uses of these materials. The Research and Development capacity in this field is strongly hampered by a lack of resources.

The Workshop therefore recommends that:
12.1. Governments and donors appreciate the need and importance of R&D and allocate funding accordingly.
12.2. State-of-the-art technologies from elsewhere should be compiled and disseminated.

12.3. Research results should be widely disseminated.

12.4. The following research issues should be addressed with priority:
- new generation of kilns,
- improvements on energy efficiency,
- use of waste as sources of fuel and pozzolana,
- appropriate production equipment,
- markets for low-cost binders,
- application technologies of alternative binders,
- properties of lesser-known alternatives, including pozzolana cowdung.

13. Coordination

It is in the interest of those involved with the enhanced dissemination of alternative binders in the East African region, that there is extensive coordination and networking between and within countries. With this in mind, the Workshop recommends that:

13.1. In view of its long standing experience and interest in alternative binders, mandate is given to ITDG to work in collaboration with interested agencies in the further exploration and implementation of the above recommendations in the region.

13.2. ITDG assumes the role of regional coordinator and will promote regional networking.

13.3. National focal points will be established at interested agencies in Kenya (ITDG), Uganda, Tanzania and Zanzibar, which will stimulate national networking.

13.4. Progress with the dissemination of alternative binders in the region should be reviewed periodically.